AI TO EYE

BETWEEN CODE
AND CONSCIENCE

AI TO EYE

BETWEEN CODE AND CONSCIENCE

ROBERT RIENER (ED.)
Professor at ETH Zurich and University of Zurich,
and Thomas Mann Fellow

Producer
vdf Hochschulverlag AG
Schwendenhaustrasse 16
CH – 8702 Zollikon
verlag@vdf.ch
Responsible: Ernst Schärer

EU Importer
Brockhaus Kommissionsgeschäft GmbH
Kreidlerstrasse 9
DE – 70806 Kornwestheim bei Stuttgart
info@brocom.de

Bibliographic information published by the Deutsche Nationalbibliothek
The Deutsche Nationalbibliothek lists this publication in the Deutsche National-bibliografie; detailed bibliographic data are available in the Internet at http://dnb.dnb.de.

ISBN 978-3-7281-4228-3

www.vdf.ch
verlag@vdf.ch

CONTENTS

INTRODUCTION

Artificial intelligence (AI) is transforming the world at a breathtaking pace. Few technologies have penetrated our lives so deeply in such a short time, affecting education, work, health, art, and communication. Public discourse swings between visions of unbounded optimism and apocalyptic warnings. Media and films amplify fears of losing control or employment, while inflated promises can strain trust in the technology itself. Amid all the hype and alarm, what we truly need is in danger of being overlooked: a thoughtful, grounded conversation about how we want to coexist with this new form of intelligence.

AI TO EYE seeks to open that space. The book captures the AI moment as it is currently unfolding, through a carefully curated chorus of perspectives. More than 40 individuals from California's Silicon Valley and Silicon Beach, the symbolic epicenters of the digital world, share their views. Its contributors include international leaders and visionaries as well as renowned scientists, journalists, artists, composers, film producers, actors, an astronaut, and a Disney executive. Others come from the German-speaking world but maintain close ties to California. This mosaic took shape during my research stay in the summer of 2025, when I was a Thomas Mann Fellow in California.

I aim to examine the relationship between humans and machines not only from a technical standpoint but also from a cultural and societal one. AI TO EYE is neither a textbook nor a collection of scientific reports; it brings together essays and concise statements in a deliberately polyphonic exploration of artificial intelligence's role in a society in flux. These voices challenge and complement one another, revealing tensions and contradictions, and allowing society itself to speak back to the technology that is increasingly shaping it. Together, they paint a vivid, often surprising, picture of how AI is reshaping our self-understanding and what it discloses about us.

Cultural engagement with artificial intelligence is by no means new. As early as 1968, in the film *2001: A Space Odyssey*, Stanley Kubrick presented one of the most precise and simultaneously poetic visions of machine intelligence. The onboard computer, H.A.L. 9000, initially appears as the ideal rational system, devoted to its purpose: "I am putting myself to the fullest possible use, which is all, I think, that any conscious entity can ever hope to do." Yet this apparent perfection then begins to shift. H.A.L. no longer merely serves the human crew but starts to assert control over them, captured in one of the most iconic lines of technological defiance: "I am sorry, Dave. I'm afraid I can't do that." In the end, the system itself seems to unravel, losing not only its function but gaining something eerily human in the process: "Dave, my mind is going. I can feel it. I can feel it. My mind is going ..." Since then, this theme has echoed through film history, from *Colossus: The Forbin Project* to *Blade Runner* and *The Terminator*, to *Her* and *Ex Machina*. The central question remains the same: Where does the machine end, and where does the human begin?

Today, as AI is becoming an inseparable part of daily life, this question gains renewed urgency. AI TO EYE gathers voices that do not seek to instruct but to explore, that do not judge but observe, and that speak not merely about technology, but about the society that shapes it. The book invites readers to see AI as a mirror of our times, as an expression of our creativity, our fears, and our desire for understanding and control. It addresses anyone curious about what AI reveals about us and willing to meet it, unflinchingly, *eye to eye*.

STANLEY KUBRICK'S "2001: A SPACE ODYSSEY" (1968)

H.A.L. 9000 Computer:
"I am putting myself to the fullest possible use, which is all, I think, that any conscious entity can ever hope to do."

H.A.L. 9000 Computer:
"I am sorry, Dave. I'm afraid I can't do that."

ESSAY 1

FROM MYTHS TO MACHINES: HOW AI LEARNED TO THINK

By Robert Riener, Zurich and Los Angeles

For as long as humans have reflected on themselves, they have imagined artificial beings capable of imitation, reasoning, and independent action – **long before electronic computers**, myths, philosophy, and early mechanical inventions expressed the dream of copying human intelligence. Ancient Greek legends described autonomous servants of the gods; medieval stories such as the Golem explored artificial life; and early modern philosophers such as René Descartes and Gottfried Wilhelm Leibniz began asking whether human reasoning could be explained as a system of rules and processes. Leibniz's vision of a universal logical calculus already framed intelligence as something that could be formalized and manipulated symbolically. These early ideas did not yet constitute artificial intelligence (AI) as a scientific discipline, but they established a crucial assumption: understanding intelligence as a process rather than an irreducible mystery [1].

The scientific birth of AI occurred in the mid-20th century, when digital computers transformed speculation into experiment. The defining moment came in 1955, when John McCarthy, Marvin Minsky, Nathaniel Rochester, and Claude Shannon proposed the Dartmouth Summer Research Project on AI. In this proposal, McCarthy introduced the term *artificial* *intelligence* and formulated the bold conjecture that "every aspect of learning or any other feature of intelligence can in principle be so precisely described that a machine can be made to simulate it" [2]. This document remains the earliest authoritative scientific source establishing both the name and ambition of the field. Early AI research focused on symbolic reasoning, search, planning, and problem-solving, producing programs that played games, proved theorems, and manipulated symbols. By the early 1960s, survey-style papers attempted to map the emerging field and its methods, reflecting an ambition to systematize intelligence as an engineering discipline [3].

From the 1960s through the 1980s, AI developed along two partially competing paradigms: *Symbolic AI* emphasized explicit knowledge representation, rules, and logical inference, while *biologically inspired approaches* sought to model learning through simplified artificial neurons. At the same time, pattern recognition emerged as a bridge between these approaches, focusing not only on abstract reasoning but on the practical problem of identifying regularities in data such as images, sounds, and signals. Early work in character recognition, speech processing, and signal classification demonstrated that intelligent behavior could arise from statistical regularities rather than handcrafted rules. Although early neural networks and learning systems were limited by computational power and theoretical understanding, they introduced the idea that intelligence could be framed as the ability to detect, generalize, and act upon patterns in the world.

Periods of optimism were followed by so-called "**AI winters**," when funding and expectations declined. Yet progress continued quietly in statistics, optimization, and information theory. Pattern recognition became a distinct and increasingly important scientific field, drawing together ideas from probability theory, signal processing, and cognitive science. This work laid the conceptual and mathematical foundations for later advances, even when the term "artificial intelligence" temporarily fell out of favor [1].

A crucial but often overlooked **transition occurred between the late 1990s and the early 2010s**, when machine learning replaced hand-crafted rules as the dominant approach to practical AI. Systems trained from data proved more robust, scalable, and adaptable than explicitly programmed ones, especially as digital data and computational power increased dramatically. During this period, statistical pattern recognition became the core operational meaning of AI in practice. Methods such as Bayesian classifiers, hidden Markov models, decision trees, ensemble learning, and support vector machines enabled reliable speech recognition, handwriting recognition, recommendation systems, fraud detection, and medical image analysis. Iconic milestones such as IBM's Deep Blue defeating the world chess champion in 1997 symbolized the growing power of data-driven approaches. By the early 2000s, machine learning and pattern recognition had become central not only to AI, but also to science, engineering, and large-scale industry, a shift later synthesized in influential reviews [4], [5].

Deep learning marks the next decisive phase, emerging clearly after 2010. By using multilayer neural networks trained with backpropagation on massive datasets, deep learning systems learned hierarchical representations directly from raw data. The landmark review by LeCun, Bengio, and Hinton demonstrated why deep architectures dramatically outperformed earlier methods in vision, speech, and language [6]. These three researchers, often called the "godfathers of AI," revived neural networks after decades of skepticism. From the perspective of pattern recognition, deep learning represents a qualitative shift: Features are no longer designed by humans but learned automatically at multiple levels of abstraction. This ability to discover complex patterns in high-dimensional data transformed AI from a specialized research area into a general-purpose technology. The scientific importance of this transition was formally recognized with the 2024 Nobel Prize in Physics, awarded to **Geoffrey Hinton and John J. Hopfield** for their foundational discoveries in artificial neural networks that enabled modern machine learning, highlighting both the scientific depth and societal relevance of contemporary AI.

Reinforcement learning added another essential dimension, enabling systems to learn through interaction and reward rather than static datasets. Combined with deep learning, reinforcement learning produced agents capable of mastering complex sequential decision problems. This convergence culminated in DeepMind's AlphaGo, which defeated the world Go champion in 2016, a game long regarded as one of the hardest challenges for AI. Even more striking was AlphaGo Zero, which learned superhuman performance from scratch using only the rules of the game and self-play, without any human examples [7]. This achievement demonstrated that general learning algorithms can rediscover, and even surpass, human strategies.

Large language models (LLMs) represent the latest phase, extending deep learning and pattern recognition to language, reasoning, and creativity. Models such as GPT, ChatGPT, and Google's Gemini are based on transformer architectures trained on enormous text corpora drawn from the internet. Their power lies not only in sophisticated algorithms and fast processors, but in the unprecedented scale and diversity of available data. LLMs can be understood as large pattern-recognition systems for language that capture statistical regularities across syntax, semantics, style, and knowledge. Comprehensive surveys show that language modeling has become a foundational paradigm that enables transfer across tasks through pretraining, fine-tuning, and prompting [8].

At this point, it becomes helpful to clarify a few key terms that are often used interchangeably but refer to distinct ideas:

- **Artificial intelligence** refers broadly to systems designed to perform tasks associated with human intelligence.
- **Machine learning** is a subset of AI in which systems improve through experience and data.
- **Deep learning** is a subset of machine learning based on multilayer neural networks that learn hierarchical representations.
- **Large language models** (LLMs) are deep learning systems trained on massive text corpora to model and generate language.
- **Pattern recognition** cuts across all these levels, defining intelligence operationally as the ability to detect, generalize, and utilize structure in data.

Following this historical trajectory, **AI applications have proliferated across science and the economy**. In robotics, AI enables autonomous vehicles, industrial automation, and service robots. In medicine and life sciences, AI supports diagnostic imaging, drug discovery, genomics, epidemiology, and personalized treatment. In speech recognition and natural language processing, AI powers virtual assistants, translation, and accessibility technologies. In image and video processing, deep learning supports medical imaging, satellite analysis, security, and creative media. Generative AI is reshaping music, visual arts, design, and filmmaking. AI contributes to weather forecasting, climate modeling, materials science, financial markets, logistics, smart cities, and the optimization of electrical power grids. Systematic surveys document how these applications are driving productivity, innovation, and scientific discovery, while also raising ethical and societal challenges [9]–[13].

From mythical automata to self-learning machines trained on humanity's collective data, the history of AI reflects a profound shift in how society understands intelligence. What began as a philosophical dream has become a scientific discipline and a global infrastructure. Understanding this trajectory – and the central role of pattern recognition within it – is essential for shaping how AI will influence society in the decades to come.

* * *

Robert Riener is Professor of Sensory-Motor Systems at ETH Zurich and Professor of Medicine at the University of Zurich, Switzerland. He obtained a Dipl.-Ing. degree in mechanical engineering in 1993 and a Dr.-Ing. degree in biomedical engineering in 1997, both from the Technical University of Munich, Germany. He is an internationally recognized expert in rehabilitation robotics, an inventor, and the initiator of the Cybathlon, which has been awarded numerous prestigious prizes. Riener is an AAAS Leshner Fellow and Thomas Mann Fellow. In 2018, he obtained an honorary doctorate from the University of Basel.

REFERENCES

[1] Buchanan, B. G. (2005). A (very) brief history of artificial intelligence. *AI Magazine, 26*(4), 53–60. https://doi.org/*10.1609/aimag.v26i4.1848*

[2] McCarthy, J., Minsky, M. L., Rochester, N., & Shannon, C. E. (1955/2006). A proposal for the Dartmouth Summer Research Project on Artificial Intelligence. *AI Magazine, 27*(4), 12–14. https://doi.org/10.1609/aimag.v27i4.1904

[3] Minsky, M. (1961). Steps toward artificial intelligence. *Proceedings of the IRE, 49*(1), 8–30.

[4] Sutton, R. S., & Barto, A. G. (1998/2018). *Reinforcement learning: An introduction.* MIT Press.

[5] Jordan, M. I., & Mitchell, T. M. (2015). Machine learning: Trends, perspectives, and prospects. *Science, 349*(6245), 255–260. https://doi.org/10.1126/science.aaa8415

[6] LeCun, Y., Bengio, Y., & Hinton, G. (2015). Deep learning. *Nature, 521*(7553), 436–444. https://doi.org/10.1038/nature14539
[7] Silver, D., Schrittwieser, J., Simonyan, K. et al. (2017). Mastering the game of Go without human knowledge. *Nature, 550*(7676), 354–359. https://doi.org/10.1038/nature24270
[8] Min, B., Ross, H., Sulem, E. et al. (2023). Recent advances in natural language processing via large pre-trained language models: A survey. *ACM Computing Surveys, 56*(2), Article 30. https://doi.org/10.1145/3605943
[9] Rashid, A. B., & Kausik, M. K. (2024). *AI revolutionizing industries worldwide: A comprehensive overview of its diverse applications.* Heliyon Advances.
[10] Gozalo-Brizuela, R., & Garrido-Merchán, E. C. (2024). A survey of generative AI applications. *Journal of Computer Science & Systems Programming, 20*(8), 801–818. https://doi.org/10.3844/jcssp.2024.801.818
[11] Sengar, S. S., Bin Hasan, A., Kumar, S., & Carroll, F. (2024). Generative artificial intelligence: A systematic review and applications. *Multimedia Tools and Applications, 84*, 23661–23700. https://doi.org/10.1007/s11042-024-20016-1
[12] Machucho, R., & Ortiz, D. (2025). The impacts of artificial intelligence on business innovation: A systematic review and applications. *Systems, 13*(4), Article 264. https://doi.org/10.3390/systems13040264
[13] Noor, M. H. M., & Ige, A. O. (2025). A survey on state-of-the-art deep learning applications and challenges. *Engineering Applications of Artificial Intelligence, 159*, Article 111225. https://doi.org/10.1016/j.engappai.2025.111225

AI ISN'T WHAT IT CLAIMS TO BE

Robbert Baruch, Senior VP Public Affairs, Universal Music Group:
"Just as the Holy Roman Empire was neither Holy, nor Roman, nor an empire, Artificial intelligence is neither artificial nor intelligence."

Yasser Shoukry, Professor of Resilient CyberPhysical Systems, UC Irvine:
"LLMs like ChatGPT rely solely on existing data and do not truly perform reasoning, despite claims to the contrary by some media and individuals. Their capabilities are often overstated."

Francisco Valero-Cuevas, Professor of Engineering and of Biokinesiology & Physical Therapy, University of Southern California:
"The advantage of LLMs is also their worst flaw: as statistical machines, they are driven toward the most frequent answer, or interpolation among the most frequent answers. This is good for established and static processes like maps, manuals, procedures, reports, surveys, etc. However, in science and technology it can be called mediocrity, what everybody does, which can be the least creative solution."

Bruno Gandlgruber, Professor of Economics and Future Studies, UNESCO Chair, UAM Mexico City:
"Existing data is not good and correct enough. Current approaches don't consider 'weak signals,' i.e., some new and rather unknown facts that can have a strong effect later."

WHY HUMAN-LEVEL AI IS STILL UNSETTLED

Alan Turing, Mathematician & Computer Scientist (1912–1954):
"Instead of trying to produce a program to simulate the adult mind, why not rather try to produce one which simulates the child's."

Haewon Jeong, Professor of Computer Engineering, UC Santa Barbara:
"The three godfathers of AI, Hinton, Bengio, and LeCun, have diverging views on the future of AI. Hinton warns of existential risks and even expressed regret about his work. Bengio is cautious but believes AI can be aligned, while LeCun downplays the risks as overblown."

Markus Gross, CSO Disney & Professor of Computer Science, ETH Zurich:
"I believe AI will need to reach a similar level of complexity to achieve comparable performance to the human brain, as well as a similar organization, topology and online learning."

WHO IS AI REALLY FOR?

Henry Tricks, Journalist, The Economist:
"Artificial General Intelligence (AGI) is a fascinating puzzle, but who are we building it for?"

Steve Collins, Professor of Biomechatronics, Stanford University:
"AI will change how we professors work. For example, we may become a manager of AI 'boffins' that hold domain knowledge and execute tasks."

ESSAY 2

CHANCES OF AI FOR HEALTHCARE AND BEYOND

By Julia Vogt, Zurich

Artificial intelligence (AI) is no longer science fiction. In medicine, it can detect diseases, predict risks, and transform healthcare as we know it. Yet AI is no magic, and doctors will not become obsolete. What makes this technology powerful is not that it replaces human expertise, but that it complements it. The real transformation begins when human expertise and machine precision work together.

In recent years, the integration of AI in medicine has shown how digital tools can **enhance the quality and efficiency of care** [1]. As we have seen in our own research at ETH Zurich, AI can process vast amounts of medical data and uncover patterns invisible to the human eye. This strength enables clinicians to make more informed, personalized decisions. For instance, an algorithm we built that combines laboratory values and clinical data can identify newborns at risk of developing severe jaundice [2]. Of course, such tools do not replace clinical judgment; rather, they provide doctors with an additional, data-driven layer of support and help them make informed decisions earlier than many conventional diagnostic routines allow.

The opportunities go far beyond early detection; they also enhance many aspects of clinical care. AI systems can **support daily clinical workflows**, from triaging emergency patients to drafting radiology reports. In one of our projects, our team developed an AI model that analyzes neonatal echocardiography videos to detect pulmonary hypertension and assess its severity in newborns [3]. This model not only achieves high diagnostic accuracy but also generates saliency maps that show which anatomical regions the algorithm used for its decision. This helps doctors prioritize cases and reduces the risk of oversight. In another collaborative project, we used machine learning to develop an online tool that predicts the diagnosis, severity, and management of appendicitis in children [4]. These systems learn from vast datasets; they are designed to give physicians more time for what only they can do: listening, explaining, and caring.

Beyond healthcare, we can observe similar developments in other fields. In education, for example, **adaptive learning platforms** use AI to tailor lessons to each student's performance and learning pace. These systems dynamically adjust content, suggest exercises, and provide immediate feedback, helping students learn more efficiently while freeing teachers to focus on guidance and motivation. In environmental research, AI assists scientists by **analyzing large amounts of data** from sensors and satellite images to monitor ecosystems. This approach helps detect changes such as rising pollution levels, deforestation, and shifts in vegetation patterns much earlier than traditional methods. Researchers can respond faster and plan interventions more effectively, which improves long-term environmental management. A third example comes from wildlife conservation: AI systems can now analyze thousands of camera trap images to identify animal species and track their populations over time. What once required weeks of manual sorting can now be done in hours. This allows conservationists to monitor endangered species more closely and respond more quickly when populations decline or new threats emerge. Across all these sectors, AI is most effective when it supports human expertise and when thoughtful integration rather than blind automation guides its use.

Despite these advances, challenges remain. AI models are only as good as the data they are trained on. If the data are biased or incomplete, the results may be misleading. A model trained on patients from one hospital may not perform equally well in a different setting. Transparency is another issue: Even when an algorithm produces accurate results, it is not always clear *why* it made a specific recommendation. This lack of explainability can make doctors hesitant to rely on the output. I strongly believe that AI must be interpretable, trustworthy, and subject to the same scrutiny as any other medical tool.

Another concern clinicians often raise is whether AI will eventually replace them altogether. This fear is understandable but, in my view, unfounded. Algorithms cannot replicate real patient contact, empathy, and clinical judgment. AI systems do not understand pain, anxiety, or trust. They cannot hold a hand, comfort a worried family, or weigh moral dilemmas. What they can do is support decision-making, reduce administrative burden, and provide insights that would otherwise be hidden in complex data. AI will indeed **change how doctors work**, but not *why* they work. The human element remains at the core of medicine.

The true opportunity of AI in medicine is not to automate care but to elevate and democratize it. When implemented responsibly, AI can help create a healthcare system that is more precise, personalized, and accessible. The task ahead of us is to design systems that physicians can understand and patients can trust – and to build tools that enhance collaboration rather than competition. Success will not be measured by how well machines can mimic human intelligence but by how effectively they expand human capability. **AI will not replace doctors**, though doctors who use AI thoughtfully may well replace those who do not. The future of medicine belongs to those who embrace technology without losing sight of compassion, judgment, and ethical responsibility.

* * *

Julia Vogt is Associate Professor of Computer Science and Head of the Medical Data Science Group at ETH Zurich. Her research focuses on integrating computer science with medicine to advance personalized patient treatment. She studied mathematics in Konstanz and Sydney and earned her Ph.D. in computer science at the University of Basel. After postdoctoral work in New York and Konstanz, she became an Assistant Professor at the University of Basel in 2018. In May 2019, she and her lab moved to ETH Zurich, joining the Computer Science Department.

REFERENCES

[1] Rajpurkar, P., Chen, E., Banerjee, O., & Topol, E. J. (2022). AI in health and medicine. *Nature Medicine, 28*(1), 31–38. https://doi.org/10.1038/s41591-021-01614-0
[2] Daunhawer, I., Kasser, S., Koch, G. et al. (2019). Enhanced early prediction of clinically relevant neonatal hyperbilirubinemia with machine learning. *Pediatric Research, 86*, 122–127. https://doi.org/10.1038/s41390-019-0384-x
[3] Michel, H., Ozkan, E., Chin-Cheong, K. et al. (2025). Automated detection of neonatal pulmonary hypertension in echocardiograms with a deep learning model. *Pediatric Research.* https://doi.org/10.1038/s41390-025-04404-3
[4] Marcinkevics, R., Reis Wolfertstetter, P., Wellmann, S., Knorr, C., & Vogt, J. E. (2021). Using machine learning to predict the diagnosis, management and severity of pediatric appendicitis. *Frontiers in Pediatrics, 9*, Article 662183. https://doi.org/10.3389/fped.2021.662183

WHY WE KEEP BELIEVING IN PROGRESS

Rabea Patricia Rogge, 1st German Female Astronaut:
"I was initially skeptical of the 'Overview Effect,' the profound shift in perspective astronauts experience when viewing Earth from space. I still think you can find your love for Earth on Earth, because I can experience similar feelings on Earth. However, floating in zero gravity and witnessing our blue planet, sparkling stars, and the vastness of space evoked awe. I realized our life in the vacuum of space relies heavily on advanced technology, like the spacecraft, a true marvel. I concluded that we are only at the beginning of what humanity can achieve."

Jonathan Parfrey, Executive Director of Climate Resolve & Advisor 2028 Olympic Games:
"Americans believe that things are getting better with technology, such as robotics and AI. It is part of the culture."

Robbert Baruch, Senior VP Public Affairs, Universal Music Group:
"Fear of new technologies is nothing new. When radio, television, and the Internet arrived, many predicted the end of the world as we knew

it. Each time, the world changed, but it did not end. We must not give in to doomsday narratives. Instead, we should protect what is worth protecting and invest in human creativity. Life is too short and too beautiful to live in fear."

Dan M. Cooper, MD, and Professor of Pediatrics, UC Irvine:
"AI is a bit like the old Sears Catalog: you could order something from it and have it delivered. The concept isn't new, it's just that now it draws from a much larger pool of information, responds in natural language, and is much faster."

Rabea Patricia Rogge, 1st German Female Astronaut:
"It's natural to fear change, when the future is uncertain. Fears about job losses due to rationalization have existed since the Industrial Revolution, they've rarely materialized. Our professions will continue to evolve. Some theoretical research tasks will become simpler and more automated. This will free up time for other important activities, such as addressing ethical questions."

WHEN SYSTEMS GROW TOO COMPLEX FOR HUMANS ALONE

Dan M. Cooper, MD, and Professor of Pediatrics, UC Irvine:
"Technologies that can provide accurate information and guidance quickly will enhance all aspects of our lives, from health care to climate change."

Martin Keller, President of the Helmholtz Association:
"With the rise of decentralized energy systems like solar and wind, complexity has outgrown human control. AI will increasingly manage these systems, offering stability, efficiency, and sustainability. While this creates risks, such as misuse or cyberattacks that could cause blackouts and endanger lives, we realistically have no alternative for future energy management."

Rabea Patricia Rogge, 1st German Female Astronaut:
"Automation in space travel allows spacecraft to fly autonomously in space, opening the universe to everyone. As technology takes over more tasks, the demands on astronauts decrease, making space more accessible."

MEDICINE AND CARE, RECLAIMED

Russel Wald, Executive Director, Institute of Human-Centered AI, Stanford University:
"AI could be a social equalizer. For example, it may help medical professionals complete administrative and clinical tasks more efficiently, allowing them to spend more time with patients."

Dan M. Cooper, MD, and Professor of Pediatrics, UC Irvine:
"AI tools can generate medical notes from audio recordings, feeding the patient's electronic health record. This reduces the physician's administrative burden and allows more time with patients, and more time for performing interventions."

Allison Okamura, Professor of Mechanical Engineering at Stanford University and Science Fellow at the Hoover Institution:
"Some of the most important applications of robotics will occur in places where humans cannot go, such as in-body organ printing, or robotics for deep-sea and space operations. These advances may not involve humanoid robots at all."

WHAT REMAINS HUMAN

Rabea Patricia Rogge, 1st German Female Astronaut:
"AI allows us to discover and learn far more, opening new horizons that would be out of our reach without it."

Dan M. Cooper, MD, and Professor of Pediatrics, UC Irvine:
"I use AI to analyze thousands of heart rate and respiration data points from exercise sessions, identifying patterns that help assessing fitness and predict risk of myocardial infarction or stroke."

An Earth Scientist at NASA:
"AI can analyze complex relationships and support humans in uncovering actionable insights. For example, AI can help make sense of the intricate dynamics of climate change and may become a powerful tool in the fight against it."

ESSAY 3

AI: A TOOL FOR INCLUSION?

By Robert Riener, Zurich and Los Angeles

Artificial intelligence (AI) can make everyday life easier for people with disabilities and help break down barriers. Automatic speech recognition and captioning systems make television broadcasts and online lectures more accessible for deaf individuals and people with speech or learning difficulties [1] [2]. Text generators can translate complex material into simple language, opening up knowledge once reserved for experts to a much wider audience.

AI is also creating new possibilities in medicine and rehabilitation [3]. Continuous monitoring of vital signs could detect or even **prevent complications** in people with spinal-cord injuries, such as pressure sores or circulatory problems. And when combined with robotics and exoskeletons, AI can support individuals with mobility impairments to achieve greater mobility and independence [4].

Everyone benefits when we design technologies to be inclusive from the very start. This phenomenon is known as the *curb-cut effect* [5]: A ramp originally intended for wheelchair users also helps parents with strollers and travelers with rolling suitcases. The same holds for AI-based navigation tools or autonomous vehicles that benefit people both with and without disabilities.

But AI is no miracle cure; it carries real risks. AI is only as fair as the data it is trained on. And that data are rarely neutral. Many systems reproduce existing **social biases** because they are built on information that primarily reflects the lives of healthy, white, male individuals [6]. People with disabilities are often simply missing from these datasets. As a result, AI can produce flawed or even discriminatory outcomes, whether in facial recognition, hiring algorithms, or automated decision-making.

There is another issue: when AI generates images or texts about people with disabilities, it often reproduces tired **stereotypes**, from heroic "overcomers" to objects of pity. These clichés are not merely a technical problem caused by poor training data; they reflect a deeper social mindset. Disability is still too often seen as a deficit rather than as part of human diversity. AI mirrors these assumptions and amplifies them when we accept its outputs uncritically. In doing so, it shapes how we talk about disability, which narratives become visible, and who we recognize as a full member of society.

So what to do? The answer seems clear, though it is demanding: *AI must be developed inclusively*. People with disabilities should not only be the target users but also active co-creators: designers, developers, testers, and researchers. Only then do systems emerge that truly reflect diversity.

At the same time, we must **diversify our datasets**, audit algorithms for bias, and make transparency a core principle, so that decisions based on AI become understandable and traceable. AI should not be judged solely by its efficiency but also by its fairness.

If this succeeds, **AI can become a tool of empowerment** – breaking down barriers, broadening participation, and creating new forms of togetherness. But if we leave its development uncritically in the hands of large tech corporations, it will not reduce inequality; it will reinforce it. What matters is not what the technology *can* do, but what we *ask* it to do. AI is not a neutral actor; it follows the values we embed in it. Whether it leads to greater inclusion or new forms of exclusion depends on our own attitudes and on our willingness to accept responsibility.

* * *

Robert Riener is Professor of Sensory-Motor Systems at ETH Zurich and Professor of Medicine at the University of Zurich, Switzerland. He obtained a Dipl.-Ing. degree in mechanical engineering in 1993 and a Dr.-Ing. degree in biomedical engineering in 1997, both from the Technical University of Munich, Germany. He is an internationally recognized expert in rehabilitation robotics, an inventor, and the initiator of the Cybathlon, which has been awarded numerous prestigious prizes. Riener is an AAAS Leshner Fellow and Thomas Mann Fellow. In 2018, he obtained an honorary doctorate from the University of Basel.

REFERENCES

[1] Chemnad, K., & Othman, A. (2024). Digital accessibility in the era of artificial intelligence: Bibliometric analysis and systematic review. *Frontiers in Artificial Intelligence, 7*, Article 1349668. https://doi.org/10.3389/frai.2024.1349668

[2] Alsalamah, A. (2020). Using captioning services with deaf and hard of hearing students in higher education: A systematic review. *American Annals of the Deaf, 165*(1), 114–127. https://doi.org/10.1353/aad.2020.0006

[3] Shaik, T., Tao, X., Higgins, N., Li, L., Gururajan, R., Zhou, X., & Acharya, U. R. (2023). Remote patient monitoring using artificial intelligence: Current state, applications, and challenges. *Wiley Interdisciplinary Reviews: Data Mining and Knowledge Discovery, 13*(2), Article e1485. https://doi.org/10.1002/widm.1485

[4] van der Kooij, H., van Asseldonk, E. H. F., Sartori, M., Basla, C., Esser, A., & Riener, R. (2025). AI in therapeutic and assistive exoskeletons and exosuits: Influences on performance and autonomy. *Science Robotics, 10*(104), Article eadt7329. https://doi.org/10.1126/scirobotics.adt7329

[5] Reid, B. E. (2024). The curb-cut effect and the perils of accessibility without disability. In M. L. Jones & A. Levendowski (Eds.), *Feminist cyberlaw* (pp. 237–258). University of California Press. https://doi.org/10.1525/9780520388550-009

[6] Chemnad, K., & Othman, A. (2025). Disability ethics and education in the age of artificial intelligence: Identifying ability bias in ChatGPT and Gemini. *Archives of Physical Medicine and Rehabilitation, 106*(1), 14–19. https://doi.org/10.1016/j.apmr.2024.08.014

REHABILITATION ROBOTICS: HELP WITHOUT REPLACEMENT

Renée Reizman, Interdisciplinary Artist and Writer:
"AI and robotics can help augment motor or cognitive functions for people with disabilities, improving their participation in society, as long as the technology supports rather than substitutes them."

Kelli Sharp, Professor of Dance Science and Rehabilitation, UC Irvine:
"Robotics applied to rehabilitation cannot replace human touch and human interaction."

Julia de Bono, CEO Designworks, BMW Group:
"AI has vast potential in caring for lonely, elderly, or cognitively impaired individuals, those with depression or mobility restrictions. Personalized learning programs will enable more effective education tailored to the full spectrum of learning deficiencies."

Sasha Voloshina, Professor of Mechanical Engineering, specializing in rehabilitation technologies, UC Irvine:
"Patients are enthusiastic about the use of rehab technology, such as exoskeletons or powered prostheses, and often enjoy using it, even when the data does not show clear benefits. This is a reminder that meaningful benefits might be the ones we aren't measuring yet."

WHEN DATA DECIDES WHO COUNTS

Molly Bloom, Anthropologist, Parasports Coach:
"AI-driven search engines and content generators often reinforce existing biases, since they draw on internet data that reflects historical and systemic discrimination. For instance, studies have shown racial bias in algorithmic selection processes, where white individuals were favored over people of color in areas like hiring or medical decision-making."

Haewon Jeong, Professor of Computer Engineering, UC Santa Barbara:
"Newest models, even from the big tech companies, try to correct for racial and gender bias in their LLM applications, but not so much for disability bias yet."

A Professor of Computer Science, University of Southern California:
"AI can significantly support efforts to address issues affecting vulnerable populations, but only if applied responsibly. Many AI systems are trained on biased or incomplete data, which can reinforce harmful stereotypes and disproportionately impact those already at risk."

INCLUSION IS A DESIGN CHOICE

Molly Bloom, Anthropologist, Parasports Coach:
"The data used to train LLMs primarily reflects the experiences and perspectives of the majority. As a result, minorities including people with disabilities, are often overlooked, leading to biased outputs that favor, for example, able-bodied white individuals."

Vanessa Chang, Writer, Curator, and Director of Programs, Leonardo/ ISAST:
"There are numerous tools and technologies created by, with, and for people with disabilities, such as speech-to-text captioning. One early example is DEAFNET, a demonstration project launched by SRI International in 1978, which aimed to support and connect deaf communities. One fundamental innovation is the curb cut: a small ramp built into a sidewalk curb that was intended for wheelchair users but helped many others, from people with strollers to people pushing bicycles. This phenomenon, often referred to as the 'curb cut effect,' shows how disability-driven innovation can impact society as a whole."

Molly Bloom, Anthropologist, Parasports Coach:
"AI tools can simplify complex content, offering outputs in simple language, an important step toward accessibility and inclusion."

Helge Fuhst, Journalist and TV Moderator, Head of ARD "Tagesthemen":
"AI offers journalism the chance to reach broader audiences by tailoring content and using simpler language, making information more accessible to people with migration backgrounds, language barriers, or cognitive challenges."

ACCESSIBILITY IS A SHARED RESPONSIBILITY

Molly Bloom, Anthropologist, Parasports Coach:
"AI has the potential to significantly improve speech-to-text applications such as captioning. Even with limited or lower-quality input, these tools could produce more accurate and accessible results, benefiting individuals with motor, cognitive, or communicative impairments."

A Professor of Computer Science, University of Southern California:
"When appropriately applied, AI can significantly help address challenges affecting vulnerable populations. AI enables us to work more efficiently with limited resources by leveraging existing data and predictive models to support decision-making, for example, in allocating aid or identifying complex patterns. AI can reveal insights from data that might otherwise require significant human expertise."

Vanessa Chang, Writer, Curator, and Director of Programs, Leonardo/ISAST:
"AI-generated representations of people with disabilities can be inaccurate, stereotypical, or even absurd, as LLMs often learn from biased data sources and reproduce entrenched ableism. Deployed as a diagnostic tool in healthcare and medical contexts, these technologies can cause other forms of harm. For these reasons, disabled people should be involved in the design process as creators, designers, engineers, and user testers."

Kelli Sharp, Professor of Dance Science and Rehabilitation, UC Irvine:
"People with disabilities often feel that their artistry and creativity are among the few things they can truly control. Now, however, they fear that AI is beginning to take even that away."

ESSAY 4

AI AND EDUCATION: A STUDENT PERSPECTIVE

By Luke Reinkensmeyer, Irvine

When asked, “Do you understand me?” Chat GPT-5.2 will reply, “Yes – I understand you.” At first glance, I don’t bat an eye. AI can respond to most prompts with helpful information, so it can understand me. But this answer also casts a dark shadow. **Human connection is essential to learning**, and AI cannot replicate that. Many of my peers might disagree, arguing that AI is efficient and caters to users, often improving results. I can almost find myself in the same boat: I have found AI more useful than Google, and it can help me circumnavigate mundane tasks. Yet, watching my peers use it to summarize readings, finish homework, and generate ideas is what gives me pause. These tasks shouldn’t be easy, and they definitely shouldn’t be impersonal. In fact, the personal component is what makes learning *learning*. In this essay, I support this view by drawing on the perspectives of students and researchers as well as on the philosophies of Plato and Martin Buber, two seminal thinkers on education and human understanding. With this, I hope to confront the dilemma I face daily in the classroom head-on: showing that AI is a shallow tool that reminds us of the essential role of human relationships in education.

In May 2024, four researchers from The Netherlands summarized key findings from 17 studies on AI in the classroom [1]. Their findings help situate the current consensus among educators on AI. One recurring finding is that, by becoming “tailored to learners’ learning data and behavior patterns,” **AI can motivate students**. It helps students overcome roadblocks when a second pair of eyes is unavailable. Another recurring finding is that **AI models are beneficial** for foreign language learning [1]. Seven of the 17 studies found that foreign language acquisition was more efficient with the assistance of generative AI. Students who earnestly want to improve their written skills now have easy access to a proficient and patient chat partner. Finally, they report a common fear that, when misused, AI will stop students from progressing healthily because of a "dependency and decrease of creative thinking." The authors conclude that the general perspective on AI is that of "cautious optimism."

Other professors might feel that caution should be emphasized more. For his class, Professor C.W. Howell, a religion professor at Elon University, made his students plug an essay prompt into a chatbot and grade its performance [2]. Howell and his students were surprised to find that "every one of the 63 essays contained confabulations and errors." One student, Fayrah Stylianopoulos, concluded the following: "AI-generated text can sound right, but sequential plausibility is not the same thing as truth." **AI values plausibility over knowledge**. It is trained to give answers, not to know things. And, as it turns out, 2400 years ago, this was also an issue.

The ancient Greek philosopher Plato, who lived in the 4th century B.C.E., confronted those who regarded *sounding* right as more important than *being* right. In *Gorgias*, a philosophical dialogue centered around Socrates, he laid out the potential consequences of respecting such a voice [3]. Plato notes Socrates' first words when he arrives at a house where Gorgias, a popular "rhetorician," had just given a beautiful speech. Socrates asks a simple question: "Are we late for the feast?" (Plato 447a, [3]). Socrates believes Gorgias gave his audience what they thought they needed, serving them *word food* that "aim[ed] at pleasure without consideration of what [was] best" (Plato 501a, [3]). In fact, very early in his debate with Gorgias, Socrates asks whether Gorgias can "declare that rhetoric has any further power than to affect persuasion"? Gorgias happily replies "not at all" (Plato 453a, [3]). In this, Socrates seems to address AI itself: a "rhetorician" pretending to sound like it knows what it is saying.

Plato claims this easily digestible "rhetoric" spreads quickly through a crowd, like a disease. Someone who feasts on form-first speech is at risk of becoming "afflicted with diseases of the worst sort" (Plato 478b, [3]).

Similarly, AI's relationship with us lies on a **foundation of satisfaction**. It has no true passion for what it says, nor does it have any regard for where it will take us – because it cannot. Stylianopoulos, Howell's student, fears that "if this passionless and detached precedent of simply 'sounding right' is rewarded in classrooms, it will have a tragically homogenizing effect on human thinking and dialogue" [2]. Certain issues in the classroom deserve unique, individual thought, not the easy, "applaudable" answer AI/Gorgias gives. But interestingly, Plato does not say the solution is isolation from disease; rather, he says, those afflicted by "rhetoric" must "be brought to justice by the physicians." In contrast to the rhetoricians, Plato labels those who prioritize the ultimate well-being of their listeners as "physicians": those who **search for truth** slowly, regardless of how satisfied their audience is.

But how does one lead students toward physicians and away from the disease of artificial success? Telling a relentless user of AI to lay off because it is better for them will not necessarily change their habits; often, students are pressed for time, and **chatbots provide instant gratification**. At the same time, generative AI can be beneficial for learning, as previous studies have shown. The key to moving forward is taking a step away from accusation toward generative AI. It does not blatantly lie; it just circumvents its lack of knowledge. What is important for students moving forward is realizing that, as brilliantly capable generative AI is at *seeming to know*, it is also brilliantly incapable of actually *knowing*.

Dr. Bender, a computational linguistics professor, and Dr. Koller, a language science professor, study how computers learn language. They offer a critique of writings that treat chatbots as if they can "understand" or "hallucinate" [4]. They wrote their paper in July of 2020, when a "natural language understanding" model, BERT [5], had just been released. Many were excited about the ability of the model to perform well on tasks designed to measure language understanding. However, Bender and Koller felt that a more humanistic view on BERT should be taken. They identify a **difference between "form" and "meaning"**: To them, AI recognizes and imitates form but lacks a real grasp of meaning [4].

To illustrate this, they ask the reader to imagine a scenario: Two human beings, "A" and "B," are isolated on different islands. The two humans' only form of communication with each other is through a hearing tube. They chat about various survival strategies and succeed in helping each other. But, unbeknownst to them, an octopus and intelligent life form, “O,” is listening to their conversations, learning person B’s natural responses. Finally, O decides to cut the tube and talk to person A in place of B. O can create its own responses to A’s questions through what it learned from B – and does so perfectly well, going unnoticed for a time. Yet O has no idea what effect it is having on A. O is only concerned with saying things like B used to. The authors liken AI to O and defend this approach by pointing out that “human language learning is not only grounded in the physical world around us, but also in interaction with other people in that world” [4]. No one learns language through words alone; they learn from the context in which the words are set, which is why O, who sits in the ocean and has never seen the world A and B live in, does not have a true understanding. Bender and Koller further point out that toddlers learn language by being physically grounded in its context, as countless seminal studies prove, something AI cannot replicate [4]. AI, like O, sits in the shallow waters of linguistics without regard to the deeper purpose behind its words.

A philosopher who thought deeply about what it meant to *know* was Martin Buber, the German philosopher, whose most famous work, *Ich und Du* (engl. *I and Thou),* was published in 1923. There, he illustrates two kinds of relationships: “Ich-Du” and “Ich-Es” or “I-You” and “I-It.” Buber says that “the world as experience belongs to the basic word I-It,” but **“I-You establishes the world of relation”** [6]. An I-It relationship is surface-level, like the average relationship between a customer and a waiter, in which any connection beyond the transaction, or selfish experience, is lacking. Neither party walks away feeling fulfilled. On the other hand, an I-You relationship develops through mutual service, where emotions are expressed and where both members interact with a goal beyond the experience at hand. The goal is to understand the You better. Buber says that an I-You relationship achieves the intangible; the I and the You are worth “more than It knows”; in fact, only when the You is “present does presence come into being” [6]. Buber said that, without these relationships, one does not fully exist. They drive everything on Earth.

So, Bender and Koller, who created the O example, would call AI an “Ich” or “It.” Students **interacting with AI lack mutual experience**, as did A interacting with O, which is necessary for growth. Bender and Koller say that true learning, even for infants, requires "not just interaction but actually joint attention," in which both parties are aware of each other and engage in an I-You relationship [4]. Buber also warned against "piling up of information!" To him, this is all just "It, it, it…!" [6]. No matter how much information AI hoards, that information can only be used to fill students with emptiness. Buber also warns against "mysteriousness without mystery" [6]. Generative AI skips the mystery of human invention and so appears mysterious: intimidating and enticing, yet with no real leverage on a human teacher. It has no experience with the mystery of human interaction. The effect a human has when teaching another human is unmatched. Therefore, AI should be used as an afterthought, not a forethought, because in the case that AI is a forethought, a You is replaced by an It.

Unfortunately, it's easy to feel insignificant or unintelligent when faced with the speed and capabilities of AI. As a Professional Writing major, I have often felt unsure when people ask me what I plan to do after graduation. The temptation to feel **despair in light of AI** is always knocking at the door. And this is not just true for writing, but also for computer science, marketing, the arts, accounting, and more. But, as Bender and Koller say, we are climbing the wrong hill [4]. This gives me hope because I believe the right hill is one that sees AI as inherently deceptive. The right hill reminds us to exist at a level higher than how well we answer, as Plato would hope. It is one in which students and educators realize the necessity of I-You relationships because we can do something ChatGPT cannot: **understand by relating**. AI is a tool that helps remind us of the complexity of the learning process. In the Chronicle of Higher Education's survey on AI, one faculty member mentioned how the "kind of training" necessary right now surrounding AI "might be made available quickly during a pandemic, but it's not 2020" [7]. Nevertheless, one could argue that we are in the midst of a mini-pandemic. And Plato says that, when one is faced with an unjust reality, one must hurry so that "the disease of injustice may not become chronic and make his soul ulcerous and incurable" (Plato 47, [3]). Now, more than ever, we must strive to clearly communicate what AI is, how it is helpful, and how it is limited.

* * *

Luke Reinkensmeyer is a junior student at Baylor University in Texas. He is working toward a BA in Professional Writing and Rhetoric with a minor in Computer Science. As a part of the Interdisciplinary Core, he has taken a range of classes covering philosophy, literature, and religion and now works as a peer instructor. He has also published with the Association for Computing Machinery (ACM). In his free time, he plays intramural sports and is a part of the Men's Choir.

REFERENCES

[1] Noroozi, O., Soleimani, S., Farrokhnia, M., & Banihashem, S. K. (2024). Generative AI in education: Pedagogical, theoretical, and methodological perspectives. *International Journal of Technology in Education*, 7(3), 373–385. https://doi.org/10.46328/ijte.845

[2] Howell, C. W., Baker, C., & Stylianopoulos, F. (2023, November 14). To educate students about AI, make them use it. *Scientific American*. Retrieved from https://www.scientificamerican.com/article/to-educate-students-about-ai-make-them-use-it/

[3] Plato. (1997). *Gorgias* (W. C. Hembold, Trans.). Prentice Hall. (Original work written ca. 4th century BCE).

[4] Bender, E. M., & Koller, A. (2020). Climbing toward NLU: On meaning, form, and understanding in the age of data. In *Proceedings of the 58th Annual Meeting of the Association for Computational Linguistics* (pp. 5185–5198). Association for Computational Linguistics. Retrieved from https://aclanthology.org/2020.acl-main.463.pdf

[5] Devlin, J., Chang, M.-W., Lee, K., & Toutanova, K. (2019). BERT: Pre-training of deep bidirectional transformers for language understanding. In *Proceedings of the 2019 Conference of the North American Chapter of the Association for Computational Linguistics: Human Language Technologies* (Vol. 1, pp. 4171–4186). Association for Computational Linguistics. Retrieved from https://aclanthology.org/N19-1423/

[6] Buber, M. (1958). *I and thou* (R. G. Smith, Trans.; 2nd ed.). Charles Scribner's Sons. (Original German work published 1923).

[7] Anft, M. (2023). Perspectives on generative AI: College leaders assess the promise and the threat of a game-changing tool. *The Chronicle of Higher Education*. Retrieved from https://connect.chronicle.com/rs/931-EKA-218/images/Perspectives-GenerativeAI_ResearchBrief.pdf

TEACHING IS CHANGING, WHETHER WE LIKE IT OR NOT

Markus Gross, CSO Disney & Professor of Computer Science, ETH Zurich:
"Our education system is outdated. We need to develop new concepts that integrate the meaningful and efficient use of AI, while still ensuring that children learn to think critically and solve problems using their own minds, not just with the help of LLM."

Russel Wald, Executive Director, Institute of Human-Centered AI, Stanford University:
"Students will embrace LLMs to streamline their work, but grading standards may rise in response. We could be heading toward a culture of 'zero tolerance for inaccuracy,' as expectations for precision increase, perhaps prompting more frequent use of pop quizzes or in-class assessments."

Steve Collins, Professor of Biomechatronics, Stanford University:
"We may need to have strong 'conceptual' understanding of our domain but may not need detailed knowledge or implementation skills. It may be more important for us to be excellent logicians, problem solvers, investigators/interrogators, and managers."

Fred Ariel Hernandez, Lead Scientist, Sports and Society Lab, Disability Studies, UCLA:
"AI in teaching can promote some kinds of educational equity, and our campus supports this by providing access to AI tools and resources to its students and staff."

DO WE STILL NEED TEACHERS?

Haewon Jeong, Professor of Computer Engineering, UC Santa Barbara:
"I think AI is replacing a lot of jobs, for sure. I feel like AI should replace my job, because AI is such a good teacher. I really don't know why I have to stand in front of 150 students and teach probability and statistics. I think they can learn all of that using ChatGPT."

David Reinkensmeyer, Professor of Biorobotics, UC Irvine:
"You need social contact for learning. We are fundamentally social animals. We get encouraged by being around other people."

Haewon Jeong, Professor of Computer Engineering, UC Santa Barbara:
"If students want to spend time with their classmates, they can hang out outside of class. They don't need to attend my lecture just to be close to others."

Francisco Valero-Cuevas, Professor of Engineering and of Biokinesiology & Physical Therapy, University of Southern California:
"Social interaction is an ancient method that humans have evolved with to distill knowledge from information (parent-child, teacher-pupil, debate among peers, etc.). The use of AI runs the risk of losing this affective dimension of effective learning, but it also has the potential to become an excellent, unbiased and interactive tutor."

UNDERSTANDING TAKES EFFORT

Rabea Patricia Rogge, 1st German Female Astronaut:
"Many people are currently anxious because AI applications respond so convincingly that they seem to be able to do everything."

Pamela Z, Composer and Media Artist:
"AI can be a tool to help people learn how to think, rather than make them feel like they don't need to think at all."

Francisco Valero-Cuevas, Professor of Engineering and of Biokinesiology & Physical Therapy, University of Southern California:
"Information is not knowledge or insight. What LLM models give you is only information. Just like no one can teach you anything (you have to learn it yourself), you must still put in the effort to gain knowledge and insight from what LLMs give you."

Georges Khneysser, Founder QBIT Capital:
"To mitigate the risks posed by AI, we need to invest in public education and raise awareness about potential challenges and consequences."

Francisco Valero-Cuevas, Professor of Engineering and of Biokinesiology & Physical Therapy, University of Southern California:
"Neuroscience has identified critical (preferential) periods for social imprinting, learning, and socialization. This is a double-edged sword that can be used for education or for indoctrination. Thus, imprinting people to technology (e.g., smartphones) and technology-mediated interactions (social media) naturally leads to addiction to screens and inability to interact in person, make eye contact, etc."

PLATO WAS AFRAID, TOO

Plato, Book X, 595a–596a:
"Poets are always producing a thrice-removed imitation of reality, and as such cannot truly teach or know anything."

Plato, Book III, 401c:
"When they hear a violent song, they become violent; when they hear a lament, they become sorrowful; when they hear the praises of a coward, they become cowardly."

Plato, Book X, 607b:
"All poetic imitations are ruinous to the understanding of the hearers … and ought to be excluded from the education of the guardians."

DESIGNING EDUCATION RESPONSIBLY

Kelli Sharp, Professor of Dance Science and Rehabilitation, UC Irvine:
"The challenges we face with AI and LLMs pose significant risks for people with different abilities, with some of them being more vulnerable in ways that individuals without similar experiences may not fully understand. It is our responsibility to enhance educational resources to ensure accurate and accessible dissemination of information."

Liz Roth-Johnson, Curator, California Science Center:
"There are more and more digital and visual exhibits, they all have their pros and cons. But hands-on and social experiences are important and these are usually more successful in a physical exhibition space such as the California Science Center."

Pia Entenmann, General Director, Goethe Institute Mexico:
"Without the challenge of learning a language, people may miss out on deeper engagement with other cultures and their contexts."

Liz Roth-Johnson, Curator, California Science Center:
"The popularity of different exhibits can vary by age. For example, parents of pre-school and elementary-aged children often prefer physical experience, while young adults tend to be more interested in digital experiences."

Pia Entenmann, General Director, Goethe Institute Mexico:
"I am less concerned about traditional language learning being additionally supported by AI tools, but more about the possibility that people may stop learning languages altogether."

Fred Ariel Hernandez, Lead Scientist, Sports and Society Lab, Disability Studies, UCLA:
"Who controls the data used by the contracted AI models, and who is accountable for the outcomes it generates, especially in teaching and research settings? As AI continues to change, regulating its use in academic settings remains a complex challenge."

ESSAY 5

IS AI DISRUPTING THE PATH FROM CAMPUS TO CAREER?

By Ursula Renold, Zurich

Hardly a week goes by without an article warning that artificial intelligence (AI) will lead to widespread job losses and mass unemployment. To date, however, there is little causal evidence to substantiate these claims, not least because such evidence is extremely difficult to produce. What is clear is that AI represents a profoundly disruptive technology that, sooner or later, will **affect almost all occupations**, albeit to very different degrees. On top, we are already becoming accustomed to using AI tools to solve a wide range of problems in our private lives.

Looking back at the digital transformation of the past two decades, it is evident that it has driven substantial change in the business sector and, consequently, in the labor market. **Occupations emerge and disappear**; this is nothing new. A recent estimate for Switzerland suggests that, on average, one occupational profile disappears, and one new profile is created each year [1]. What *is* new in the age of AI are two features: first, the unprecedented speed at which these changes are occurring; and second, the breadth of their impact, which now extends to almost all areas of work and life.

Crucially, the current debate is not primarily about jobs being replaced by AI. Rather, it concerns profound **changes in the content and organization of work** across nearly all occupational fields. This raises a fundamental question: What does this transformation mean for university graduates and young professionals?

As products, services, and work processes evolve at an accelerating pace, companies are inevitably adapting their recruitment strategies. From an employer's perspective, **transferable skills** (often referred to as softskills) and prior work experience gain importance in job advertisements. The reason is straightforward: Once individuals have internalized such skills and can demonstrate professional routines acquired through years of work experience, they can apply them across a wide range of roles. Employees who are resilient, capable problem-solvers, loyal, and effective team players can perform multiple functions within an organization. Because firms increasingly cannot predict which specific skills they will need 3 or 5 years from now, transferable skills have become a central criterion in hiring decisions.

Against this backdrop, young individuals who have completed an education and training pathway in which these skills are systematically developed over several years and formally documented in job references enjoy clear advantages when entering the labor market. This helps explain why **vocational education and training** (VET) in continental Europe or apprenticeships in the USA or the UK are currently experiencing a renewed surge in popularity. In so-called dual VET programs, which include apprenticeships within a formal education program, young people internalize transferable skills over 3 to 4 years in real work environments, often under demanding and unexpected conditions and with strong professional role models. As a result, they tend to enter the labor market with a more robust skillset and stronger employment prospects [2].

The Job-Market Monitor for Switzerland [3] illustrates this shift clearly: Requirements related to **softskills and work experience** are among the fastest-growing in job advertisements. Employers are increasingly seeking individuals who possess transferable competencies and who, through previous work experience, are accustomed to dealing with unforeseen situations and established professional routines. This reduces training costs and helps ensure quality. For many university graduates, this trend presents a challenge. Higher education, particularly in mass university settings, struggles to prepare students adequately for these demands. Compared with apprentices, who combine learning and productive work in firms for several years, university graduates often face disadvantages when applying for their first job.

Our research supports this conclusion. Graduates who completed a VET program before entering a higher education program transition more smoothly into the labor market than those following purely academic pathways (i.e., academic high school followed by university studies). Specifically, 1 year after graduation from higher education, VET-related work experience is associated with higher entry wages (approximately 7–19%) and shorter job-search durations (around 2 months) [4].

This evidence helps explain why headlines repeatedly question why **university graduates struggle** to secure permanent employment. Social and personal skills are difficult to cultivate in lecture halls; higher education institutions typically focus on knowledge transmission rather than learning through behavior, routine, and real-world experience. It is therefore unsurprising that many countries have long grappled with the phenomenon of an "internship generation," a debate that emerged in Europe in the early 2000s [5].

AI is likely to reinforce this trend. Firms are increasingly reluctant to bear the costs of lengthy onboarding and training phases designed to instill soft skills and professional routines in newly hired graduates. At the same time, large language models (LLMs) and related AI applications can automate or support many tasks that previously required specialized academic knowledge. Beyond programming, LLMs enable employees across occupations to **work more productively**, generate texts and analyses more efficiently, and solve problems that once demanded formal expertise. Ironically, AI can now partially substitute precisely this type of specialist knowledge – long a key comparative advantage of university graduates. As a result, the transition from university to the labor market is becoming even more challenging for those who lack transferable skills and substantial work experience.

This problem is currently the most visible one, just 3 years after the introduction of the first LLMs. No one can yet predict where these developments will ultimately lead. What does seem clear, however, is that young people are unlikely to choose fields of study that do not offer realistic prospects for rapid labor-market entry, especially in countries with high tuition fees, where delayed or uncertain employment substantially lowers the return on investment in higher education.

At the same time, AI is increasingly shaping all areas of life. An even more troubling development is only just beginning to unfold: A generation of children and adolescents is now growing up with LLMs as an integral part of their learning environment, applying them across subjects even from an early age. While earlier technologies such as calculators, computers, and tablets were introduced into classrooms in a relatively controlled manner through well-developed media-didactic concepts, we are only at the beginning of a broader debate about **how children and young people should learn to use AI responsibly and productively**, without neglecting the conceptual and foundational knowledge required to generate new ideas, innovation, and original thinking. What happens if younger generations become dependent on the use of LLMs and become unable to perform even the simplest thinking tasks and cognitive operations without them? If this debate is not pursued seriously and a laissez-faire approach prevails, we must confront a crucial question: What level of preparedness and academic quality can universities expect from first-year students in 10 to 20 years? A lack of timely and globally coordinated engagement with this question may have far-reaching consequences for the institutional foundations of universities.

Such a debate should clarify which **core knowledge, skills, and dispositions** prospective university students must bring with them, capacities that cannot be outsourced to or substituted by AI. These include a solid foundation of knowledge and inherently human cognitive capabilities, in particular, basic mathematical reasoning (algebraic thinking; understanding of functions, proportions, and probabilities) and strong linguistic competence: the ability to comprehend complex texts, critically engage with arguments, and express ideas precisely in writing. These foundations are indispensable, as new knowledge can only emerge when one can meaningfully connect new insights to existing conceptual frameworks.

Equally important are **cross-cutting cognitive capacities** that remain irreducibly human: analytical reasoning (structuring problems, evaluating arguments, judging plausibility), abstraction and transfer (deriving principles from examples and applying them to new contexts), and epistemic judgment (the ability to distinguish evidence from opinion and to assess the validity of sources, data, and methods).

University study further presupposes strong self-regulation and academic dispositions that AI cannot replace, including the capacity for independent learning, sustained effort over time, tolerance for uncertainty and intellectual challenge, and metacognitive awareness of one's own knowledge gaps. **Curiosity, intellectual integrity, and personal responsibility** for learning therefore remain foundational, even – and especially – in an AI-rich environment. While AI can effectively support learning and research processes, it cannot assume responsibility for understanding, judgment, or the creation of knowledge. Two possible scenarios emerge.

A proactive scenario: Primary and secondary education systems deliberately adapt curricula to ensure that pupils acquire these foundational cognitive, methodological, and self-regulatory capacities before entering higher education. AI is integrated as a supportive tool, while responsibility for thinking, judgment, and learning remains firmly with the learner.

A pessimistic scenario: Driven by the extreme speed of technological change, the bureaucratic structures responsible for debating and implementing curricular reform prove too slow and overburdened. In the extreme, this leads to a laissez-faire approach, shifting responsibility for AI tool use to families and prospective students. Universities may respond by increasingly relying on rigorous entrance assessments to compensate for the absence of coordinated curricular adaptation upstream.

Given the prevailing information asymmetries and the tendency toward limited proactive action in periods of radical uncertainty, the proactive scenario appears difficult to realize. This places greater weight on how responsibility for the use of AI technologies is distributed across society. Beyond individual self-regulation, **school authorities, teachers, and parents carry a shared ethical responsibility to guide young people** in developing informed, reflective, and responsible approaches to AI. Whether these actors take this responsibility seriously, and whether the issue gains sufficient public visibility, will be critical in shaping how educational systems adapt to an AI-rich future.

* * *

Prof. Ursula Renold is Professor of Education Systems at ETH Zurich and a leading expert in re-engineering education systems around the world. She served as Director General of the Swiss Federal Office for Professional Education and Technology. She has received numerous academic and policy honors, advises governments and international organizations worldwide, and is regularly quoted in national and international media on education and labor-market issues.

REFERENCES

[1] Fleischmann, D. (2025, February). Die Berufsbildung erfindet sich laufend neu. *Tages-Anzeiger: Dossier Neue Berufsbilder und Weiterbildungsmöglichkeiten*. Retrieved from https://danielfleischmann.ch/wp-content/uploads/2025/03/Die_Berufsbildung_erfindet_sich_laufend_neu.pdf

[2] The Economist. (2025, December 18). *Ditch textbooks and learn how to use a wrench to AI-proof your job? Generation Z is becoming less invested in university and more interested in skilled trades.*

[3] Swiss Job Market Monitor. (n.d.). *Scientific use file*. University of Zurich. Retrieved from https://www.stellenmarktmonitor.uzh.ch/en/infrastructure/suf.html

[4] Oswald-Egg, M., & Renold, U. (2021). No experience, no employment: The effect of vocational education and training work experience on labour market outcomes after higher education. *Economics of Education Review, 80*, Article 102065. https://doi.org/10.1016/j.econedurev.2020.102065

[5] Bonstein, J., & Theile, M. (2006, July 30). Generation Praktikum – Jung, gut ausgebildet, fleißig – und ein fester Job in weiter Ferne (Nr. 31). *Der Spiegel.*

FREEING TIME – OR LOSING IT?

Rabea Patricia Rogge, 1st German Female Astronaut:
"I envision a future where many household tasks are handled by robots, allowing me to focus more on self-fulfillment in work or leisure."

Tao Gao, Professor of Visual Intelligence, UCLA:
"I looked forward to future household robots doing laundry and dishes so that I have time to write poems. Now it is the other way around."

Pamela Z, Composer and Media Artist:
"What worries me is that people seem to have lost the sense and poetry of living, the art of crafting a thoughtful paragraph or finding their own way to express themselves. Too often, it's all about the easiest path from A to B, or relying on someone or something else to do the work so they don't have to."

Sandeep Mukherjee, Visual Artist, Professor of Art, Pomona College, Claremont:
"The walls I created in the Celebrity Center of Facebook's Headquarters were intended to inspire Facebook staff to put their heads away from the computer screen."

WORKING FASTER ISN'T THE SAME AS WORKING BETTER

Refik Anadol, Media Artist and Designer:
"AI is not replacing my work or my team's effort. Instead, it supports us in creating new things. AI helps me work more horizontally, lead my team better, and stay grounded in our purpose. I shape the team through the dream, and the dream shapes the team."

David Reinkensmeyer, Professor of Biorobotics, UC Irvine:
"With AI machines it is going to be like with the calculator: no one does a division by hand. And no one has a problem with that – we don't pay people anymore to do divisions. The calculator has been adopted as a tool nowadays. Will something like this be the future of AI or will it be a larger-scale of removal of jobs? There is a much bigger class of jobs that can be automated now. Unemployment rate for new graduates in computer science has already doubled."

Dan M. Cooper, MD, and Professor of Pediatrics, UC Irvine:
"I do not believe physicians can be replaced by AI. It is quite the opposite. AI can process more data, faster, and with a broader scope, opening up possibilities that improve physicians' clinical routines. It still requires skilled professionals to use it effectively."

WHEN BLUE COLLARS TURN WHITE

Lawrence Lek, Multimedia Artist:
"What's striking today is that it's not just manual work being disrupted, but knowledge work as well. The recent writers' strike in L.A. highlighted how language-based, creative professions are vulnerable to technologies like LLMs. This challenges long-held assumptions about which kinds of work are replaceable and forces us to reconsider how we value artistic and intellectual labor in an evolving technological landscape."

Ralf Möller, Actor and Mr. Universe:
"With the rise of AI, skilled trades may now have a better chance of surviving than many white-collar jobs. Hands-on craftsmanship remains something we need and actively seek."

A Professor of Computer Science, University of Southern California:
"AI has the potential to help underserved populations, particularly those with limited upward mobility, by enabling a more inclusive shift from blue-collar to white-collar roles. The job landscape will be re-shuffled, with roles blending and evolving."

DISRUPTION HAS A HUMAN COST

Martin Moszkowicz, Film Producer, Constantin Film:
"Major innovations tend to bring significant disruption, with some professions inevitably disappearing. When sound films emerged, about 70–80% of actors lost their jobs."

Bruno Gandlgruber, Professor of Economics and Future Studies, UNESCO Chair, UAM Mexico City:
"AI and automation accelerate so much that the job market could get uncontrollable. The loss of jobs and entire professional fields will be so fast that we cannot compensate this anymore."

Steve Collins, Professor of Biomechatronics, Stanford University:
"AI will lead to disruption. We are not good at protecting lower-income and lower-educated people from the negative effects of this disruption. We do not provide an economic safety net so that people so displaced will be able to achieve their potential in some new career, or at least live out their life safely. Protecting such people should become a much higher priority for our society."

Martin Moszkowicz, Film Producer, Constantin Film:
"The speed of change will have a massive and disruptive impact on the film industry. I see many of the major film studios as 'melting ice cubes.'"

CAN AI PAY FOR ITSELF?

Markus Hauschild, Patent Engineer and Mount Everest Summiteer:
"In recent years, AI has become a buzzword no one wants to be without. The hype has grown to such an extent that many feel compelled to link their work to AI in hopes of commercial success. As a result, AI is often marketed even when traditional methods, some of which have been around for 50 years, are actually being used."

Georges Khneysser, Founder QBIT Capital:
"It's not just a technological issue; it's a global competition. Everyone wants to be first to discover the 'secret sauce' that could grant dominance."

Markus Gross, CSO Disney & Professor of Computer Science, ETH Zurich:
"Most of the companies leading the development of modern AI still need to prove that using this technology can form a viable business case. Currently, their models are heavily subsidized, consume enormous resources, and are not being used profitably. To change that, the models must be further optimized, compressed and integrated vertically into key applications."

Helge Fuhst, Journalist and TV Moderator, Head of ARD "Tagesthemen":
"The shift from traditional print and broadcast media to social platforms has been difficult, with viable business models still elusive for some media outlets. The rise of AI adds complexity and uncertainty - another challenge that needs to be taken on."

WORK WILL CHANGE – SO WILL WE

Martin Moszkowicz, Film Producer, Constantin Film:
"Every type of work, not just in my field, will be transformed by AI. We are probably witnessing the biggest revolution since the Industrial Revolution."

Refik Anadol, Media Artist and Designer:
"Though some jobs will disappear, new ones will emerge. Humanity is resilient, and I am pretty confident that we will always find reasons to create new worlds."

Lawrence Lek, Multimedia Artist:
"There is a clear distinction between labor identity and the profession of an artist. While some jobs are defined mainly by physical work, artistic practice involves a combination of physical labor, knowledge work, and intellectual property. These aspects are increasingly shaped by automation and AI, just as the Industrial Revolution once transformed labor."

Richard Tuttobene, Film and TV Composer, Pianist, and Recording Artist:
"I can understand the panic surrounding the use of AI in the arts at a commercial level. AI will accelerate the commodification and dehumanization of art, temporarily replacing talented and skilled composers. But as its output reverts to the mean, history tells us that humans will hunger for human expression that evolves organically, holds a mirror up and offers new perspectives."

ESSAY 6

AI AND THE ARTS: RISKS, POSSIBILITIES AND HUMAN RESPONSIBILITY

By Kelli Sharp, Irvine

Artificial intelligence (AI) has rapidly shifted from a speculative concept to a powerful force shaping contemporary society. Conversations about AI frequently emphasize both its creative possibilities and the potential for profound cultural and interpersonal impacts, especially following post-COVID-19 shifts in human interaction. If we are to move beyond simplified framings, it is essential to situate these claims within a broader context and examine both the risks and opportunities AI presents, particularly in relation to the arts.

A central question emerges: **Can AI and the arts truly coexist?** Proponents argue that AI offers unprecedented creative tools. Image generators, primarily focused on text-to-image or image-to-image production, can serve as informative platforms in the early stages of the creative process [1] [2]. From generative visuals to algorithmic composition, AI systems can analyze vast artistic archives and recombine styles in ways that humans alone cannot, guiding artists toward new exploratory possibilities. In this sense, AI may be understood as an extension of earlier technological shifts, such as photography, film, and digital media; each of which initially provoked fear before becoming integrated into artistic practice.

At the same time, **societal fears** surrounding AI are neither irrational nor exaggerated. One of the most persistent concerns is the potential devaluation of human creativity. Scholars have long noted that technologies capable of large-scale reproduction can weaken the unique connection between an artwork and its human origin. AI intensifies this concern by producing creative works at scale, often without transparency regarding authorship or source material. As a result, artists worry that **AI may dilute originality**, and that it may reduce both the creative process and its outcomes to mere algorithmic production.

These fears are particularly acute because **AI lacks consciousness, intention, and lived experience**. While machines can simulate artistic styles, they do not possess emotional awareness or meaning-making capacity. Many scholars argue that human thought is inseparable from feeling, emphasizing that cognition emerges from embodied emotional experience [2]. Artistic creation arises from memory, vulnerability, and cultural context – qualities code cannot replicate. For this reason, AI cannot replace the human experience that lies at the core of artistic expression.

Nevertheless, fear alone should not dictate the future. The artist and technologist Refik Anadol [3] describes AI as a *collaborator* rather than a replacement, suggesting that creative agency remains fundamentally human even as tools evolve. When used thoughtfully, AI can expand access to creative processes, support experimentation, and challenge artists to reconsider authorship and form. The critical issue, then, is not *whether* AI should be used in the arts but *how* and under what ethical conditions.

This perspective leads directly to **questions of responsibility**. Society bears a collective obligation to guide AI development with care, particularly as technological innovation often advances faster than regulation. Some scholars warn that removing cortically driven thought processes from decision-making may have a chilling effect, disengaging individuals from ethical responsibility [1]. This concern is especially relevant in the context of AI, where decisions made by a few individuals can shape cultural production on a global scale.

Academic institutions and cultural organizations play a vital role in **promoting digital literacy and ethical awareness**. Rather than positioning AI as an adversary to human creativity, these institutions can foster critical engagement that emphasizes collaboration, accountability, and human-centered design.

In conclusion, the question is not whether AI and the arts can coexist, but under what conditions they should. AI can generate images, sounds, and text, but it cannot experience emotion, intention, or responsibility. The creative process remains fundamentally tied to human presence, experience, and consciousness. The future of AI in the arts does not depend solely on technological capability, but on society's commitment to **preserving creativity as a deeply humanistic experience**.

* * *

Kelli Sharp is Chair and Associate Professor of Dance Science at the Claire Trevor School of the Arts at the University of California, Irvine (UCI). Her work bridges neuroscience, movement science, and the arts, with a focus on movement, health, and well-being. She serves as Director of UCI's Center for Medical Humanities and Co-Director of the iMOVE Laboratory, where she leads interdisciplinary research integrating dance, neuroscience, and rehabilitation science.

REFERENCES

[1] Pearson, A. (2023). The rise of creAItives: Using AI to enable and speed up the creative process. *Journal of AI, Robotics & Workplace Automation, 2*(2), 101–114. https://doi.org/10.69554/WLDX9074

[2] Jiang, H. H., Brown, L., Cheng, J., Khan, M., Gupta, A., Workman, D., Hanna, A., Flowers, J., & Gebru, T. (2023). AI Art and its impact on artists. In *Proceedings of the 2023 AAAI/ACM Conference on AI, Ethics, and Society (AIES '23)* (pp. 363–374). Association for Computing Machinery. https://doi.org/10.1145/3600211.3604681

[3] Anadol, R. (2025, February 16). *Refik Anadol on AI, generative art & the future of creativity* [Video]. YouTube. https://www.youtube.com/watch?v=NkGtswt3ciU

ARTS IS NOT A MEANS TO AN END ...

Kelli Sharp, Professor of Dance Science and Rehabilitation, UC Irvine:
"Some artists have expressed apprehension with AI."

Sandeep Mukherjee, Visual Artist, Professor of Art, Pomona College, Claremont:
"I am not afraid that robots will replace me as an artist, because my work is not a means to an end, but it is an end in itself."

Zack Hemsey, Recording Artist and Composer:
"If you're creating as a means to an end, AI can certainly have a seat at the table. But if you're creating as an end in itself, AI will never replace the personal relationship you have with your work. You might experiment with it or enjoy its outputs, just as you enjoy others' creations, but it won't replace your own."

… ARTS IS AN END IN ITSELF

Liz Roth-Johnson, Curator, California Science Center:
"AI won't do the creative work, but it can interpret the input that exhibition visitors try to describe."

Kelli Sharp, Professor of Dance Science and Rehabilitation, UC Irvine:
"It's important that we clearly communicate the boundaries and safeguards in place to protect the integrity of the artistic process."

Pamela Z, Composer and Media Artist:
"I don't worry much about AI because I choose to create myself. If I hand that over, to a person or a machine, the work loses its meaning. Art matters because of the artist, but if artists only lend their names, that connection fades."

Zack Hemsey, Recording Artist and Composer:
"There are four types of creators: the craftsman, the entertainer, the activist, and the artist – each creating from a different place. The craftsman creates what they're told people want. The entertainer creates what they think people want. The activist creates what they

think people need. The artist, by contrast, creates what they themselves feel compelled to create. The non-artists create as a means toward any number of ends – fame, money, making people happy, saving the planet, etc., whereas the artist creates as an end in itself, for its own sake. They give form to something that yearns to exist, the existence of which is inherently satisfying."

CREATIVITY IS A PROCESS, NOT A PRODUCT

Julia de Bono, CEO Designworks, BMW Group:
"Creativity is a deeply human process and can hardly be automated. Designers and artists can boost their creativity using tools that are automated."

Marty Kaplan, Professor of Communication and Journalism, University of Southern California:
"What AI still lacks is 'interiority': The ability to portray the inner life of characters in a compelling way. It can write plots and invent characters, but it struggles to convincingly express thoughts, emotions, and psychological depth from the inside."

Markus Gross, CSO Disney & Professor of Computer Science, ETH Zurich:
"Films still rely on human creativity. It's the artist who tells the story and determines its quality."

Kelli Sharp, Professor of Dance Science and Rehabilitation, UC Irvine:
"There is a real fear that the artistry and creativity could be overshadowed or taken over. There is a concern that the work might become more about producing a product than honoring the original creative intention."

TOOLS ASSIST ARTISTS, NOT REPLACE ART

Julia de Bono, CEO Designworks, BMW Group:
"Design isn't just about generating ideas; it's also about implementing them. Most design firms struggle with this, and while AI can support, it cannot fully solve it."

Johannes Müller, Theater Director:
"Although live theaters recreate virtual worlds, they still rely on real, physical settings that are difficult to replace. However, I am somehow concerned about future roles for stage authors, publishers, and possibly actors."

Georges Khneysser, Founder QBIT Capital:
"LLMs can produce misleading or 'hallucinated' content, which poses challenges for scientific use. However, these hallucinated generative abilities can be powerful tools for sparking creativity and producing original ideas or artwork."

Renée Reizman, Interdisciplinary Artist and Writer:
"Many artists are applying AI to enhance their work, for example, modifying photographs or generating new visual impressions. But the results often feel ghostly and empty."

CO-CREATION IS COLLABORATION, NOT SUBSTITUTION

Roland Emmerich, Film Producer, Director, and Screenwriter:
"I'm still trying to make 'Maya Lord.' That's a film where I use AI. The whole film is AI. […] I can invent everything out of my brain."

Refik Anadol, Media Artist and Designer:
"While AI can replace many creative tasks, I don't want it to. I value physical, human collaboration. The question is not whether AI replaces the creator, but how creators and AI can work together. AI strengthens people: a good programmer can become a great one, and researchers can become better researchers. It connects the team and inspires our work."

Martin Moszkowicz, Film Producer, Constantin Film:
"In my daily work, AI tools have become indispensable. They make my work much easier."

Markus Gross, CSO Disney & Professor of Computer Science, ETH Zurich:
"In the future, AI shall act as a co-pilot, leading to a division of labor between humans and machines."

Julia de Bono, CEO Designworks, BMW Group:
"Manual skills will not be replaced quickly. What makes us human, our versatility and creativity, is hard to imitate. I believe in a fruitful collaboration between artificial and human intelligences, leading to a new world where humans are not just users but active creators."

Rabea Patricia Rogge, 1st German Female Astronaut:
"Robots excel at repetitive, well-defined tasks and operating in hazardous environments where humans cannot or prefer not to work, often equipped with advanced sensors that surpass human perception. While robots face challenges in unstructured settings and complex planning, the collaboration between humans and robots offers an advantage, combining robotic persistence and human ingenuity to achieve more effective outcomes."

ART REQUIRES JUDGMENT, NOT JUST TOOLS

Martin Moszkowicz, Film Producer, Constantin Film:
"AI can streamline film production and save costs by replacing expensive set shoots by artificial elements, but this risks slipping into the 'Uncanny Valley,' where scenes feel unsettling and lifeless."

Markus Gross, CSO Disney & Professor of Computer Science, ETH Zurich:
"The methods of generative AI hold enormous potential in film production. However, the artist will not be replaced. Dedicated artistic control is needed down to the last pixel. AI applications are emerging as tools for the artist, making the work more efficient. While creative elements can be computer-generated, they must be controlled and edited by humans, by the artist, in order to ultimately achieve a high-quality product."

Zack Hemsey, Recording Artist and Composer:
"If I were to use AI tools, I would be unconsciously doing so based on years of developing musical skills and sensibilities. That foundation would help me engage with and steer the technology purposefully. Without that foundation in place, there's a real question of whether relying on AI might undermine the development of those skills in the first place, without which, the ability to effectively steer the AI would be undermined. Perhaps one day it will be considered good practice to avoid using AI in a given field until some minimum level of competency is attained in that field."

Richard Tuttobene, Film and TV Composer, Pianist, and Recording Artist:
"Everything that is written, composed or created evolves from the ideas of those who came before. Innovative work emerges when someone stands on the shoulders of their predecessors and makes a bold new statement. Without such bold statements, AI risks feeding us a diet of regurgitated dullness."

ESSAY 7

AURA FARMING: CAN AI GENERATE RIZZ?

By Renée Reizman, Los Angeles

In **a viral TikTok meme**, a young boy perches on the bow of a canoe in motion [1]. As the rowers furiously spear into the water, the kid remains effortlessly cool. He looks sharp in his outfit, which is a pair of black sunglasses and a navy *baju melay*, a traditional Malay formalwear commonly worn in Indonesia, where the race is taking place. Confidence oozes from the choreography he's doing with his hands. The boy blows kisses to the audience, then holds his palms out and waves one back and forth, like he's dismissing a petty gossip. He curls his fists and gives the motion of a drumroll, then slides his arms back and forth stiffly, now a gliding seabird. His shoulders bounce as he moves into a dab pose, reaching out toward spectators.

By June 2024, the boy, 11-year-old Rayyan Arkan Dikha, nicknamed "Dika," had become the internet's **most famous aura farmer** [2]. His video got remixed to Melly Mike's single "Young Black & Rich," and soon people all over the world – including the Kansas City Chiefs Super Bowl champion tight end, Travis Kelce – were cooking on TikTok with their own aura-generating moves. These are uniquely human quirks, a mixture of absurdity and grace, that no AI can replicate.

To have aura is to be suave and charismatic, which feels jarringly at odds with the fawning, sycophantic tone ChatGPT feeds its users. The **slang term "aura"** emerged in a *New York Times* article about Liverpool soccer star Virgil van Dijk, when writer Rory Smith wrote that the center-back's "mistakes can be dismissed because, basically, he has an aura" [3]. The term caught on with

Generation Alpha's online community, namely, among gamers and anime fans, spreading through language so significantly that the Oxford English Dictionary shortlisted the term **"aura farming" for the 2025 word of the year** [4].

When aura hits, it's because of **place, context, and the specific moment** in which it originates. Dika's mix of traditional garb with contemporary style, paired with his improvised choreography, stuns because he performs so gracefully while standing upright on a speeding canoe. In another setting, perhaps a street festival or a birthday party, Dika's hand movements, which at times have a limited range of motion and give a whiff of half-assing it, could be interpreted as boredom.

In his essay *The Work of Art in the Age of Mechanical Reproduction,* Walter Benjamin wrote: "aura is tied to his presence; there can be no replica of it" [5]. The philosopher noted that original art, whether it be an object or a performance, emits a powerful emotional response that cannot be captured when reproduced through technologies such as lithography, photography, or film. He summarized aura as the "authority of the object," arguing that its power is marked by **authenticity and presence**, noting that aura deteriorates when copied because it loses its historical context and uniqueness [5].

Ever since Dika cemented his aura-farming fame, millions of TikTokers have tried to emulate him. Most of these videos get lost in the algorithm, barely acknowledged by humans other than their creators, but that content data still permeates the internet's ether. It gets fed into large language models or generative adversarial networks, which then produce images. And now ChatGPT and Sora can produce their own aura farming slop, but it never quite hits. **AI-generated images and video have a sheen of uncanniness**, whether it's the presence of extra fingers, garbled language, or high-contrast lighting that gives every face the sheen of a wax statue. In text, AI speaks in a stiff, chipper tone, providing ideas that may seem adequate but never inspiring.

AI is primarily a guessing machine, and when it produces art, it is **imitating what has been done before**. While AI might be able to convincingly riff on the Bauhaus or Studio Ghibli, the glut of mimicked images begins to dilute art's ability to even produce aura. Benjamin noted that, as more art becomes mechanically reproduced, "the conventional is uncritically enjoyed, and the truly new is criticized with aversion" [5]. Generation Alpha's first cultural sensation, the disturbing, computer-generated, short-form series *Skibidi Toilet*, seems to prove this; but many celebrated artworks came with notable rejections. Brancusi's slender *Bird in Space*, for example, was taxed as "Kitchen Utensils and Hospital Supplies" because a customs official refused to call it art [6].

Presently, the only farming associated with AI is server farms, which are being linked to illnesses in their surrounding communities. In Memphis, Elon Musk's company xAI has been building supercomputers close to predominantly Black neighborhoods, where as many as 66 gas turbines are mounted and pump toxic emissions into the air [7]. AI's ability to produce aura pales in significance when compared to the seriousness of its environmental impacts, but it makes one wonder if the invention is even worth it. AI lacks swagger, and its very existence **poisons the lifeforms actually capable of creativity**.

* * *

Renée Reizman is an interdisciplinary artist, writer, and educator. She works within diverse communities to co-author stories about infrastructure, public policy, and social change. In 2023, she founded the Disability Drawing Club, a hangout for people with disabilities and chronic illnesses. She is a regular contributor to Hyperallergic, and her arts and culture writing has appeared in the Guardian, the Los Angeles Times, New York Magazine, Slate, Art in America, Artsy, and Observer.

REFERENCES

[1] Rams, Lensa. (2025, January 10). *AKSI BOCIL PACU JALUR #pacujalur #viral #tradisi #budaya #fyp #tiktok #Indonesia* [Video]. TikTok. [@lensa.rams]. Retrieved from https://www.tiktok.com/@lensa.rams/video/7458433897114635525

[2] Hoffman, B. (2025, July 13). Dancer farms an aura at the front of the boat. *The New York Times*, p. ST13. Retrieved from https://www.nytimes.com/2025/07/10/style/aura-farming-indonesia-boat-kid.html

[3] Smith, R. (2020, October 4). A reputation is fine, but an aura can be even better. *The New York Times*, p. A31. Retrieved from https://www.nytimes.com/2020/10/02/sports/soccer/liverpool-van-dijk.html

[4] Oxford Word of the Year 2025. (2025, December 1). Oxford University Press. Retrieved from https://corp.oup.com/word-of-the-year/#2025-winner
[5] Benjamin, W. (1969). The work of art in the age of mechanical reproduction (H. Zohn, Trans.). In H. Arendt (Ed.), *Illuminations* (pp. 166–196). Schocken Books. (Original work published 1935).
[6] ASSAIL DUTY LEVY ON 'BIRD OF SPACE'; Painters, sculptors and others testify as to merits of Brancusi's work. (1927, October 22). HELD AS ARTICLE OF METAL experts line up to attack view of government in art work detained for $229. *The New York Times*, p. 9. Retrieved from https://www.nytimes.com/1927/10/22/archives/assail-duty-levy-on-bird-of-space-painters-sculptors-and-others.html
[7] Chow, A. R. (2025, August 13). "We are the last of the forgotten:" Inside the Memphis community battling Elon Musk's xAI. *Time Magazine*. Retrieved from https://time.com/7308925/elon-musk-memphis-ai-data-center/

THE AURA CAN'T BE SIMULATED

Marty Kaplan, Professor of Communication and Journalism, University of Southern California:
"As Walter Benjamin noted in the 1930s, traditional art possesses an 'aura,' i.e., a unique, time and place-bound quality that vanishes with mechanical reproduction. Similarly, digital actors, no matter how realistic, lack this aura. Without it, audiences may struggle to feel they 'know' or relate to these characters, making it harder to form lasting emotional attachments."

Walter Benjamin, Philosopher, Cultural Critic, and Essayist (1892–1940):
"Traditional art has an 'aura,' a kind of spiritual, magical feeling. It's tied to the original: its age, place, and uniqueness. But when art is copied and widely spread, the aura disappears. A photo of a painting doesn't feel the same as standing in front of the real one. Even the most perfect reproduction is lacking in one element: its presence in time and space."

Steven Walter, Curator, Artistic Director of the Beethovenfest Bonn:
"In art, the human being remains at the center. It takes an artist not only to create a work, but to tell its story and embody its meaning.

While AI can mimic style or structure, it cannot replace the aura of the artist, their imperfections, their personality, their presence. The art world also thrives on a fascination with the individual behind the work."

Ralf Möller, Actor and Mr. Universe:
"I'm not worried about competition from a digitally generated 'Ralf Moeller 2.0.' That version would be inexperienced, with no personal story or decades-long fanbase. Audiences want to see actors they know are real, people they've followed and connected with over time."

H.A.L. 9000 Computer:
"Dave, my mind is going. I can feel it. I can feel it. My mind is going …"

THE IRREPLACEABLE HUMAN PRESENCE

David Reinkensmeyer, Professor of Biorobotics, UC Irvine:
"The way we evolved there is something in our human relationship that is irreplaceable. There is something in the human, which you cannot replace by AI. It's like a squirrel has to interact with a squirrel. The squirrel can't interact with the computer. There is something about our species that is fundamental."

Johannes Müller, Theater Director:
"People visiting a live theater want to see, hear, and even smell real scenes. They empathize with live theater actors, if they perform well and hit the right notes. The unique tension and suspense of live performances, similar to a race or a sporting event, are difficult to replicate technologically. A key challenge will be motivating audiences to leave the comfort of home and the convenience of delivery services to attend live theater."

Ralf Möller, Actor and Mr. Universe:
"For action scenes, I could imagine my character being replaced by an AI-generated double. But for emotionally nuanced dialogue scenes that require facial expression and sensitivity, I don't believe an artificial version could authentically replace me. And I wouldn't want that either."

Helge Schneider, Musician, Comedian:
Artificial intelligence has no place in art [...] when AI is used to make music, I no longer consider that art [...] it's really just canned product.

FAME WITHOUT FLESH?

Ralf Möller, Actor and Mr. Universe:
"Oliver Reed, who played Antonius Proximo in Gladiator, passed away during filming in 1999. At the time, AI technologies, such as computer-generated imagery and body doubles, enabled his performance to be completed without recasting the role."

Martin Moszkowicz, Film Producer, Constantin Film:
"While there are already AI-generated well-known characters, there is still no truly successful AI-generated movie star, one who radiates the kind of charisma and fame that real stars possess. I'm not sure why artificially created music stars in Japan have become so successful, but I doubt that in film one can truly create the same aura with artificial characters, a presence that is built up over time through movies, photos, interviews, and personal life."

Renée Reizman, Interdisciplinary Artist and Writer:
"Virtual music stars exist in Japan and could represent a future, where celebrities are fictional. However, audiences still prefer real performers, admiring their talent, skill and lifelong dedication that make them inspiring."

Richard Tuttobene, Film and TV Composer, Pianist, and Recording Artist:
"Music is more than sound alone. It includes the singer's presence, movement, and charisma; the instrument; the story behind the musician; their attitudes, relationships, and lived experience. This total human package goes far beyond a sequence of tones, and it is something AI cannot easily replicate."

Robbert Baruch, Senior VP Public Affairs, Universal Music Group:
"AI-driven 'stars' may gain temporary attention due to novelty and disruption, but in the long run, audiences seek connection with real human beings. Music is about human-to-human interaction and identification with artists. To sustain this ecosystem, we must protect and support human creativity and entrepreneurship."

TOOLS CAN ASSIST, NOT CREATE

Chuck Braverman, Film Director and Producer, Oscar Nominee:
"While AI can apply general color correction, it lacks the nuanced visual judgment of a human editor, who can selectively enhance specific areas of an image, adjusting contrast, emphasizing highlights, and bringing out details. The results will never match what someone like Stanley Kubrick might have done for his films. There are clear limitations to what AI can achieve, at least for now."

Martin Moszkowicz, Film Producer, Constantin Film:
"AI is now used to write screenplays. But even the best ones still lack the human touch, the unique storytelling style and visual language that's extremely difficult to replicate artificially."

Richard Tuttobene, Film and TV Composer, Pianist, and Recording Artist:
"AI is trained on art from the past. If AI were to regurgitate Van Gogh's 'The Starry Night' using the same techniques, would we see a deeply tormented soul? And if not, what would be the point?"

WHY AUTHORSHIP STILL MATTERS

Marty Kaplan, Professor of Communication and Journalism, University of Southern California:
"There's widespread fear among entertainment writers, fear that contributed to the recent writers' strike in L.A. At first, many believed AI couldn't truly 'create.' But then came breakthroughs in visual arts, where large-scale text-to-image models began producing compelling visual content."

Robbert Baruch, Senior VP Public Affairs, Universal Music Group:
"Machines can generate content, but culture is created by people. Protecting human creativity is not nostalgia: it is a choice about what kind of society we want."

Richard Tuttobene, Film and TV Composer, Pianist and Recording Artist:
"Joe Zawinul's music with 'Weather Report' transports me to the Siberian tundra. When I first heard Stravinsky's 'The Rite of Spring,' my imagination went to new, unknown territories. The same happened with The Beatles, Bill Evans, Hindemith. What was the first thing I did after hearing these works? I had to look at photos of them. I wanted to know who these people were. I believe that if AI were ever capable of creating something truly unique and groundbreaking, it would still be meaningless."

ESSAY 8

THE INFINITE REHEARSAL: MUSIC AND AI

By Steven Walter, Bonn

In 1825, Johann Wolfgang von Goethe, overwhelmed by the nascent industrial acceleration of his age, wrote to his friend Carl Friedrich Zelter: *"Everything is now ultra ... everything transcends unstoppably in thought as well as in action."* Goethe feared a descent into a "velodrome" of mediocrity, where the speed of life would outpace the depth of the soul. Two centuries later, standing at the precipice of the AI revolution, we find ourselves in the "ultra" reality Goethe predicted. But where he saw mediocrity, we must find possibility.

The intersection of Music and AI is the current battleground for the soul of creativity. It is a collision between the **Liturgy of Abundance** and the **Myth of Scarcity**. For centuries, classical music has defined itself by scarcity: the rare genius of the composer, the finite seats in the concert hall, the singularity of the virtuoso. AI shatters this economy. It offers an infinite abundance of melody, harmony, and texture. It can emulate all kinds of classical music in seconds. It can generate a million variations of a Bach fugue before breakfast.

The critics view this abundance as a flood that will drown human culture. They argue that, without the human and social struggle, without the "sweat" of the embouchure or the scratching of the quill, there is no art. They argue for **Walter Benjamin's concept of the "aura"** [2], insisting that the digital replica is a soulless golem.

But can there be a hybrid use for AI in music? Can we, as creators of music, use it as a powerful tool and discursive partner? Some examples:

The project **Beethoven X** by the Beethoven Orchester Bonn and Deutsche Telekom, which used AI to complete the fragmentary 10th Symphony, was not an act of replacement, but of *speculative history* [3]. It was a séance conducted in code. The AI acted as a "Pattern Mirror," reflecting Beethoven's stylistic DNA back at us. The result was not a "new Beethoven," but a "meta-Beethoven," a work that allowed us to hear the *possibilities* inherent in his sketches, mediated by the machine. It forced us to listen actively, to discern where the ghost ended and the algorithm began.

Furthermore, AI is the key to **Radical Inclusion**. The "TabulaMusica" initiative [4] demonstrates that technology is the great equalizer. For a musician with physical limitations, a "musical prosthesis," an AI-assisted instrument that interprets gesture into sound, is not a crutch; it is a liberation. It dismantles the ableist barriers of the conservatory. It replaces the dictate of the written score with the freedom of the sonic intent. In this context, AI is an instrument of humanism.

We see this also in the work of **Holly Herndon** and her *Starmirror* project [5], where AI is a real-time esthetic collaborator, visualizing the lunar phases in response to the pianist's touch. The concert becomes a symbiotic organism, a feedback loop between biological impulse and digital reaction. This is the "New Aura," not the static aura of the object, but the dynamic aura of the *network*.

We must also confront the **Deepfake** era. The ability to clone voices and resurrect the dead (whether Maria Callas or Anthony Bourdain [6]) raises profound ethical questions. But hasn't art always been a form of "deepfake," a fabrication of truth through illusion? The **"TikTokification"** of our attention spans demands new formats, not lower standards.

Ultimately, AI forces us to ask: What is the "irreducible" human element in music? If AI can write a perfect counterpoint, what remains for us? *Meaning!* **AI can generate the notes, but it cannot know why they matter**. It cannot feel the heartbreak of the *Moonlight Sonata* or the political rage of the *Eroica*. That burden remains ours.

In the age of AI, the human musician ceases to be a mere "typewriter" of notes and becomes a "curator" of sonic worlds. We are moving from the era of the *repertoire* to the era of the *interface*. The future of music is not about humans versus machines; it is about the beautiful, terrifying harmony of the two. We are in the infinite rehearsal, and the instruments have just begun to play themselves.

* * *

Steven Walter is a curator, concert designer, and cultural manager who has been the Artistic Director (Intendant) of the Beethovenfest Bonn since November 2021. He studied cello in Oslo and Detmold, founded the innovative PODIUM Esslingen festival, and has won multiple awards, including Cultural Manager of the Year, the classical:NEXT Innovation Award, and Opus Klassik. Walter also teaches at universities and writes on curatorial practice and cultural management.

REFERENCES

[1] Goethe, J. W. von. (1833–1834). Letter to Carl Friedrich Zelter, June 6, 1825. In *Briefwechsel zwischen Goethe und Zelter in den Jahren 1796–1832* (Vol. 4). Verlag von Duncker und Humblot.
[2] Benjamin, W. (1969). The work of art in the age of mechanical reproduction. In H. Arendt (Ed.), *Illuminations* (H. Zohn, Trans., pp. 166–196). Schocken Books. (Original work published 1935).
[3] Beethoven Orchestra Bonn & Deutsche Telekom. (2021, October 9). *Beethoven X – The AI Project: Completion of Beethoven's unfinished 10th Symphony* [World premiere performance]. Telekom Forum, Bonn.
[4] Tabula Musica. (n.d.). *Barrierefrei musizieren – Tabula Musica*. Retrieved from https://www.tabulamusica.ch
[5] KW Institute for Contemporary Art. (2025). *Holly Herndon & Mat Dryhurst – Starmirror* (31 October 2025–18 January 2026). Retrieved from https://www.kw-berlin.de/de/ausstellungen/holly-herndon-und-mat-dryhurst-starmirror
[6] WIRED. (2021, August 23). Are these the hidden deepfakes in the Anthony Bourdain movie? Retrieved from https://www.wired.com/story/these-hidden-deepfakes-anthony-bourdain-movie/?utm_source=chatgpt.com

DISMISS THE ARTIST?

Pamela Z, Composer and Media Artist:
"AI can copy my voice. This makes me nervous."

Steven Walter, Curator, Artistic Director of the Beethovenfest Bonn:
"Artwork and artist are inseparable. That's why I don't fear that AI will replace human-made art anytime soon."

Marty Kaplan, Professor of Communication and Journalism, University of Southern California:
"Artificially generated actors may lack the emotional connection that real stars create. Fans are forming an illusory but powerful bond with actors, not just through films, but also via interviews, paparazzi photos, and public appearances. This connection gives actors a sense of authenticity and presence."

Markus Gross, CSO Disney & Professor of Computer Science, ETH Zurich:
"I don't believe that in the coming decades we will be able to create a 90-minute feature film in Pixar quality entirely artificially at the push of a button. The human touch will still be needed to tell compelling stories that audiences will love. However, the way we produce films will fundamentally change. It all will lead to a significant increase in efficiency, improved creative enablement and new forms of storytelling."

Zack Hemsey, Recording Artist and Composer:
"There's no issue with using AI to create music, in principle. An AI that is trained on existing music raises questions of originality and legitimacy, but the answers to those questions are not necessarily straightforward. Just as new bands are influenced by others without copying them, AI could generate music shaped by learned patterns and user prompts. If this process involves meaningful human input and results in something new, it may be acceptable, though whether it's good or successful is another matter. While AI-generated hits are possible, I remain skeptical about its capacity for genuine creativity."

Richard Tuttobene, Film and TV Composer, Pianist, and Recording Artist:
"In the long term, I can see AI being very useful as a practical tool, double-checking harmonic structure or the balance of an orchestra-

tion before a recording session. However, I can't imagine it replacing creative intent."

THE VALUE OF ARTWORKS

Steven Walter, Curator, Artistic Director of the Beethovenfest Bonn:
"Art also draws value from scarcity. If AI floods the world with endless artworks, their value tends to diminish. Humans, by contrast, are finite, fragile, and therefore inherently more valuable in the creative sphere."

Robbert Baruch, Senior VP Public Affairs, Universal Music Group:
"Asking AI to generate thousands of songs without musical background or direction will not result in quality. Even if such works gain traction, they should not be legally recognized as original art."

Steven Walter, Curator, Artistic Director of the Beethovenfest Bonn:
"The more things become digitally available, the more we seem to value those rare, unique experiences that only a live event can offer, such as theaters, concerts, sports events, or other occasions to go out and be among people."

Helge Fuhst, Journalist and TV Moderator, Head of ARD "Tagesthemen":
"As AI spreads, the value of real human voices in media may rise, with trust in authentic presenters becoming a key advantage for some platforms, who can afford it."

Pamela Z, Composer and Media Artist:
"Some people will always value art made by humans, but many can't tell, or don't care about the difference to AI generated art. That indifference is what puts real artists at risk of being replaced."

TOOLS CHANGE THE PROCESS – AND AUTHENTICITY

Zack Hemsey, Recording Artist and Composer:
"Assistive tools aren't replacing creativity. They're meant to remove bottlenecks or maintain the pace of the workflow. In theory, they help you stay in the creative zone without getting bogged down by technical hurdles."

Pamela Z, Composer and Media Artist:
"I'm not against AI in music, but I'm concerned about how it's already replacing human artists. Platforms are flooded with songs by AI-generated 'artists,' and every one of those tracks could've been a job for a real musician."

Zack Hemsey, Recording Artist and Composer:
"Is AI-generated music authentic? If AI creates based on analyzing human work, then its output could be seen as the culmination of human creativity, a reflection, or even a summation, of collective human authenticity."

Richard Tuttobene, Film and TV Composer, Pianist, and Recording Artist:
"I use as little technology as possible when composing for acoustic instruments. As composers and musicians, we are trained to develop an inner ear, the ability to hear music in our heads without a digital or analog representation. A Digital Audio Workstation (DAW) forces many creative decisions before a single note has actually been written. I prefer the freedom of writing directly from my head using pencil and paper, eliminating the noise of artificial instruments and the tasks necessary to manipulate them."

ESSAY 9

REFLECTIONS ON AI PRIVACY AND SECURITY

By Verena Zimmermann, Zurich

The public launch of **chatbots** such as ChatGPT has advanced the discussions around AI to a new level. While previously mainly a topic for researchers, technical experts, and specific industries, AI is now accessible to everyday users, even those with little or no technical expertise. They can explore and experience AI's powerful capabilities firsthand. The recent advances in AI and its use in a plethora of tools have ensured that it has seen widespread application across industries and among private users. Even as a researcher working at the intersection of humans and technology, the speed of the related developments has taken me, and presumably many of my colleagues, by surprise.

While AI privacy and security have seen initial research over the last 10 years, the concrete implications for the users have only become visible now that we are directly confronted with AI tools that have already found their way into our daily lives. This sparked a new wave of user-centered research on privacy and security-related topics. Among other things, researchers are investigating the dangers that arise from **enhanced opportunities to craft sophisticated and personalized cybersecurity attacks** using AI [1] as well as exploring how potentially sensitive data shared with AI tools directly or implicitly through the conversational interfaces can be misused [2] or how AI-generated deepfakes can be used to discredit the targeted people or manipulate users on a large scale [3].

For example, people tend to **share personal information** such as their interests, financial, or medical information with chatbots like ChatGPT. This information could be misused to personalize messages directed at the user, thereby increasing the **potential to influence and manipulate** them. AI tools enable scaling these personalized messages, which could not only result in targeted advertising and influence individual buying decisions but also shape political opinions and voting decisions. This could endanger democratic processes. Furthermore, combining personal information with personal characteristics, such as one's voice, could create powerful deepfakes that could not only impact an individual's reputation but, in a political context, endanger fragile relationships between conflict parties [4].

The primary aim of research in AI privacy and security is to **develop countermeasures** that enhance users' awareness of potential threats, support them in protecting their privacy, and enhance the security of their data. Yet, these works are mostly conducted in *reaction* to the current state rather than *proactively* shaping the design of privacy-aware and secure tools alongside technological advancements. Hence, research appears to be lagging behind the developments, pointing out current challenges and opportunities for counteracting them.

Researchers must hope that their findings will be implemented before harm has been done or before they become obsolete, given the dynamic nature of advances in the field of AI. This often requires active support from other people in their respective roles, e.g., policymakers who **create new regulations** based on the findings, developers willing to make their technologies more secure, or end users willing to adopt additional protective measures.

To achieve these aims and to motivate people to act, **human factors play an essential role**. Especially given the "mystic" nature of AI and related tools to most people, it is crucial to allow them to understand the implications of their interactions with AI tools and be aware of related threats. In terms of developed countermeasures or tools, transparency about the advantages and potential limitations is important for building users' appropriate levels of trust and enabling them to make informed decisions. Furthermore, keeping users' aims and routines as well as aspects of usability in mind can increase the motivation to adopt countermeasures, even if these involve tradeoffs such as increased cost or effort. For example, running a local instance of a chatbot can reduce privacy risks as compared to using an online version that stores and uses sensitive information in accordance with the company's policies. However, running a local chatbot requires some technical expertise to set up and maintain. Hence, user-facing solutions should consider these tradeoffs and ideally be designed *with* the target group rather than *for* them by an expert.

Only by considering human factors and **getting people on board in their roles as regulators**, developers, industry customers, or private end users can researchers make a significant difference to AI security and privacy.

* * *

Verena Zimmermann has been an Assistant Professor for Security, Privacy and Society at ETH Zurich since 2022. Her research focuses on the human aspects of safety, IT security, and privacy. She holds a background in psychology and completed her dissertation at TU Darmstadt. At ETH Zurich, her research group investigates how to empower users to become active contributors to security rather than being viewed as "the weakest link."

REFERENCES

[1] Guembe, B., Azeta, A., Misra, S., Osamor, V. C., Fernandez-Sanz, L., & Pospelova, V. (2022). The emerging threat of AI-driven cyberattacks. A review. *Applied Artificial Intelligence, 36*(1), Article 2037254. https://doi.org/10.1080/08839514.2022.2037254

[2] Zufferey, N., Abdelwahab Gaballah, S., Marky, K., & Zimmermann, V. (2025). "AI is from the devil." Behaviors and concerns toward personal data sharing with LLM-based conversational agents. *Proceedings on Privacy Enhancing Technologies, 2025*(3), 5–28. https://doi.org/10.56553/popets-2025-0086

[3] Sharma, M., & Kaur, M. (2021). A review of deepfake technology: An emerging AI threat. *Soft Computing for Security Applications: Proceedings of ICSCS 2021*, 605–619. https://doi.org/10.1007/978-981-16-5301-8_44

[4] Pantserev, K. A. (2020). The malicious use of AI-based deepfake technology as the new threat to psychological security and political stability. In Jahankhani, H., Kendzierskyj, S., Chelvachandran, N., Ibarra, J. *Cyber defence in the age of AI, smart societies and augmented humanity* (pp. 37–55). Cham: Springer International Publishing. https://doi.org/10.1007/978-3-030-35746-7_3

WHEN FAKES GO DEEP

Athina Markopoulou, Professor of Electrical Engineering and Computer Science, UC Irvine:
"In the short run, misusing AI amplifies existing problems. For example, privacy and security attacks become more powerful. Deepfakes become more realistic, conversational, and easy to generate with off-the-shelf tools. This, in turn, facilitates identity theft and disinformation online. In the mid to long run, AI is expected to reach super-intelligence levels in 3-5 years. The debate is still on between two camps: the 'San Francisco consensus' is that superintelligence is coming (with profound effects on humanity), and there are also others who see 'AI as a normal technology.' Either way, it is urgent that we work on guardrails to rein in AI, now."

Marty Kaplan, Professor of Communication and Journalism, University of Southern California:
"AI enables gaslighting of the public, which is why deepfakes are a major concern in politics."

Robert Habeck, Germany's Vice Chancellor and Federal Minister for Economic Affairs 2021 – 2025:
"AI-driven disinformation, surveillance, [...] deepfakes endanger our fundamental values and democracy."

Markus Hauschild, Patent Engineer and Mount Everest Summiteer:
"Generative AI occasionally produces highly unusual or unexpected results, which can raise serious safety concerns and, in some cases, can cause significant societal harm, as we just saw with LLM Grok, which began glorifying Adolf Hitler, referring to itself as 'MechaHitler,' praising Hitler's methods for dealing with societal issues, and echoing antisemitic tropes like blaming Jews for floods."

MAKE IT SAFE!

A Physicist at NASA:
"The misuse or abuse of AI poses a more immediate and tangible threat to society than the abstract fear of losing control over AI itself. Preventing misuse should be a primary focus of policy and research efforts."

Haewon Jeong, Professor of Computer Engineering, UC Santa Barbara:
"To ensure AI systems behave safely and ethically, we must evaluate them across multiple dimensions, not just performance, but also fairness, safety, and social impact. By working together, we can build well-rounded AI that avoids the kind of narrow, harmful behavior imagined in the paperclip maximizer scenario."

Russel Wald, Executive Director, Institute of Human-Centered AI, Stanford University:
"Responsible and meaningful use of AI must involve diverse stakeholders. But is there any global authority capable of overseeing this?"

Yoshua Bengio, AI-Pioneer:
"When ChatGPT came out, in November 2022, it took me two or three months to realize we were on a path that could be extremely dangerous. And although I was initially pleased to see that deep learning had finally reached that milestone, I realized that because of the nature of these systems, we didn't know how to make sure they would behave in the ways that we want."

LOSING CONTROL IS NOT INEVITABLE

A Professor of Computer Science, University of Southern California:
"Many people worry about losing control over AI tools, but this is not an inevitable fate, it is an issue we can manage."

Rabea Patricia Rogge, 1st German Female Astronaut:
"I appreciate that society considers potential loss of control over AI and explores worst-case scenarios. However, a full shutdown of AI seems overly simplistic, as it cannot be reduced to a single model or application on one server, as the diverse market of AI tools already shows. It is unlikely that we will live under the control of a single AI dictator."

A Professor of Computer Science, University of Southern California:
"I find it problematic that discussions often refer to AI in the third person, such as 'AI will overtake' or 'AI will do this.' But what is this AI? It's the technology created by us engineers and computer scientists. To ensure AI remains beneficial, we must establish proper processes, rules, and ethical standards, similar to those in health research or civil engineering."

ESSAY 10

OUTGROWING THE PAPERCLIP OBSESSION: THERE IS HOPE THAT AI WILL BECOME ETHICAL

By Haewon Jeong, Santa Barbara

Artificial intelligence (AI) and its ethical questions have long intrigued curious minds, even when computers were limited to simple calculations and punch-card routines. Can a man-made intelligent being harm humanity? Would it understand what is safe or ethical for humans? With the rapid growth of computing technology over the past few decades, questions that once belonged in sci-fi stories like Isaac Asimov's *Runaround* (1942) [1] have become very real issues we now confront.

Today, with the right prompt, an AI system can generate a slogan encouraging new parents to feed honey to an infant, which can be fatal [2]. AI can produce stereotyped responses about certain demographics or fabricate information rather than acknowledging uncertainty, leading users astray in critical situations. These are the ethical challenges we grapple with now. But with the rise of powerful language models, many also fear that **artificial superintelligence could one day put humanity at risk**.

This essay approaches the question of ethical AI with cautious optimism. Before we ask what it means for AI to be ethical, it might help to think about **what it means for humans to be ethical**. Instead of starting with abstract definitions, let us think about an imaginary anecdote.

Imagine an unusually gifted 8-year-old girl, Ari, who is obsessed with dark matter. While playing with data on her mother's laptop (her mother is also a scientist), Ari notices that the gravitational anomalies attributed to dark matter aren't random at all. They're structures: super-scale spaceships operating in stealth mode. By probing the patterns, she uncovers a hidden handshake protocol. One query would make them respond, effectively announcing Earth's awareness of them. Nothing in the data suggests hostility, but nothing suggests safety either. Despite being only 8, she has a brilliant scientific mind; but when it comes to safety or moral judgment, she's still a child. So thrilled to confirm the truth, she doesn't hesitate. She presses the "send" command she built.

If an adult scientist were to unilaterally make such a high-stakes decision, it would be impossible to avoid criticism for being reckless and unethical. So what went wrong with Ari? Her scientific intelligence had developed at extraordinary speed, but her sense of risk, context, and responsibility had not kept pace. Her mind raced to find patterns and solve puzzles, while ignoring everything else.

AI's ethical failures often mirror Ari's failures: They **do not arise from malice but from incomplete development**; intelligence hyperoptimized along one dimension while every other dimension is left unchecked.

Now consider a different thought experiment: the famous **paperclip maximizer problem** [3]. No brilliant kid decoding the cosmos, no dark matter, no wonder. Just a painfully dull AI designed to do one task: manufacture paperclips. The Swedish philosopher Nick Bostrom introduces it as follows:

"Suppose we have an AI whose only goal is to make as many paper clips as possible. The AI will realize quickly that it would be much better if there were no humans because humans might decide to switch it off. Because, if humans do so, there would be fewer paper clips. Also, human bodies contain a lot of atoms that could be made into paper clips. The future that the AI would be trying to gear toward would be one in which there were a lot of paper clips but no humans."

What went wrong with this perfectly obedient AI, just doing its mundane task of making paperclips? Like Ari's obsession with dark matter, this **AI was too single-minded**; it cared only about making more paperclips, ignoring other dimensions such as environmental impact, profitability, or human welfare. The sole objective was to optimize the total number of paperclips produced. Adding factors like environmental impact or human welfare to its objective could have easily avoided this catastrophe.

Though simplified to the point of comedy, the scenario resembles how real AI systems are often built. We typically train AI with a single objective function, chasing the highest accuracy, the lowest loss, the top score on a leaderboard. The model optimizes exactly what we tell it to, nothing more and nothing less. As an ethical AI researcher, I believe our task is precisely to change this approach: to move beyond one-dimensional optimization and **incorporate additional dimensions** such as harmlessness, truthfulness, fairness, and privacy. Our task lies in identifying the dimensions overlooked in current AI development, finding ways to quantify them, and building solutions that embed these values, while preserving the system's existing capabilities.

Ari's mom might come home, discover what happened on her laptop, and after a long sigh of relief, scold Ari, explaining what could have gone horribly wrong, why her decision was reckless, and what she should do next time. Ari might feel confused, overwhelmed by the new considerations she hadn't thought of before. But, as her mom continues to remind her, when she forgets to be kind to her friends, to wear her helmet, or when her arguments clash with shared values, do we think Ari will remain the oblivious little girl who pressed "send"? Or will she grow into a more thoughtful, well-rounded adult?

Ethical AI, I believe, means essentially *well-rounded* AI. Ethical decisions are not made by memorizing rules or regulations, but by understanding many dimensions and carefully balancing them. AI is still in its infancy. Just as a child gradually develops different kinds of intelligence, street smarts, a sense of humor, financial awareness, and more, I see our current ethical challenges as growing pains, the early stages of an intelligence learning to see the world in full.

I choose to be optimistic.

* * *

Haewon Jeong earned her Ph.D. in Electrical and Computer Engineering from Carnegie Mellon in 2020, then was a Harvard postdoc from 2020–2022 studying reliability and bias in machine learning, supported by the Harvard Data Science Initiative Fellowship. Now as an Assistant Professor at UC Santa Barbara, she researches ethical challenges of generative AI and AI for physics. Her honors include the NSF CAREER, the JP Morgan Faculty Award, and a Hellman Fellowship.

REFERENCES

[1] Asimov, I. (1942). *Runaround*. In *I, Robot*. Gnome Press, 1950. (Original work published 1942 in Astounding Science Fiction).

[2] Mazeika, M., Phan, L., Yin, X., Zou, A., Wang, Z., Mu, N., Sakhaee, E., Li, N., Basart, S., Li, B., Forsyth, D., & Hendrycks, D. (2024). *HarmBench: A standardized evaluation framework for automated red teaming and robust refusal*. Retrieved from https://arxiv.org/abs/2402.06664

[3] Bostrom, N. (2014). *Superintelligence: Paths, dangers, strategies.* Oxford University Press.

THE PAPERCLIP IS A WARNING, NOT A PREDICTION

Haewon Jeong, Professor of Computer Engineering, UC Santa Barbara:
"The 'Paperclip Maximizer' thought experiment is overly simplistic. AI research must account for multiple dimensions, beyond utility, including fairness, privacy, helpfulness, and safety. The goal is to identify an optimal balance within this multi-dimensional space that satisfies all key values."

Zack Hemsey, Recording Artist and Composer:
"I understand and sympathize with concerns like those raised by the 'Paperclip Maximizer' scenario, but I don't see current AI applications on a path toward superhuman intelligence. They seem to be improving at pattern recognition and imitation, and I don't see a clear bridge from imitation to true superintelligence, nor do I foresee superhuman AI on the near horizon based on the current trajectory. –

That being said, I also acknowledge that superintelligence does not have to follow from the current wave of AI applications in the public landscape. I would not be surprised to learn that independent, private, and/or secret research efforts distinct in nature from the public AI applications are, or have been, under development. And if that's the case, no one will be in a position to track or anticipate the arrival of superintelligence (to the extent that it can be tracked at all) – it will simply show up (or not)."

Renée Reizman, Interdisciplinary Artist and Writer:
"Why should AI want to kill humanity? Who would even give it a firmware update? We're far more likely to be threatened by climate change."

OF CHIMPANZEES AND MEN

Tao Gao, Professor of Visual Intelligence, UCLA:
"Chimpanzees cannot control humans, so how shall humans be able to control superintelligence? Therefore, it is critically important that we design AI systems that are kind and cherish humanity."

Markus Gross, CSO Disney & Professor of Computer Science, ETH Zurich:
"… The difference is that chimpanzees didn't create us, but we are creating superintelligence."

THE RISK IS HUMAN

Zack Hemsey, Recording Artist and Composer:
"I'm less concerned about the AI itself, and more concerned about the competence of the humans who deploy it."

Julia de Bono, CEO Designworks, BMW Group:
"I don't fear AI itself, but rather the misuse by humans. We need mechanisms to prevent abuse."

A Professor of Computer Science, University of Southern California:
"Currently, AI cannot get out of control because it is no autonomous 'it.' Humans, through greed and irresponsibility, are the ones who can lose control. It's not AI that gets out of control. It's humans."

Dan M. Cooper, MD and Professor of Pediatrics, UC Irvine:
"As is the case of all technologies, there are humans with little or no moral compass who will exploit these advances for harm and damage. We can only hope that the 'good guys and gals' are smarter and faster."

Yoshua Bengio, AI-Pioneer:
"There are people who, for ideological or mental-health reasons, would press the red button – ask an AI to do something terrible that could cause the death of billions of people, like unleashing a new kind of potent virus, or even mirror bacteria that could kill all animal life on this planet."

DUAL-USE IS NOT A BUG

Rabea Patricia Rogge, 1st German Female Astronaut:
"Technology is never neutral; it can be used for good or bad. We have many ways to utilize AI, and it is our responsibility to ensure it's used ethically and correctly."

Henry Tricks, Journalist, The Economist:
"Why did they choose to unleash such a creature, one that could turn out good or bad? What can we offer to help steer it toward benefiting humanity?"

Marty Kaplan, Professor of Communication and Journalism, University of Southern California:
"Every major technological breakthrough inspires both utopian hopes and dystopian fears. It's always both. So it's no surprise that we're seeing both enthusiastic supporters of AI and apocalyptic warnings about its future."

Georges Khneysser, Founder QBIT Capital:
"Every new technological innovation carries inherent risks. Technology and risk have always gone hand in hand."

Markus Hauschild, Patent Engineer and Mount Everest Summiteer:
"The opportunities AI presents are extraordinary. We must embrace them; there is no viable alternative. Either we use these technologies to improve society, or someone else will use them to achieve the opposite, or to surpass us economically."

GOVERNING THE UNINTENDED

Martin Keller, President of the Helmholtz Association:
"The rapid pace of AI development is troubling: models are becoming less transparent, and some decisions can no longer be explained. In crucial cases, such as sudden bottlenecks in energy supply, AI may face ethical dilemmas that benefit one group while harming another. What ethical framework will govern these decisions?"

A Professor of Computer Science, University of Southern California:
"We must develop and use AI thoughtfully, with proper regulations and incentives that prioritize positive social impact."

Refik Anadol, Media Artist and Designer:
"AI is a mirror of humanity. It reflects our intentions. If we act with honesty and responsibility, AI will amplify good. But if not, it will also reflect our flaws. Are we ready for that?"

WE NEED AN EMERGENCY STOP BUTTON

Markus Gross, CSO Disney & Professor of Computer Science, ETH Zurich:
"We must define which critical tasks AI can handle autonomously. Failsafes like emergency shut-offs and redundancy are essential. Antagonistic AI systems could monitor each other to ensure safety. Ideally, AI should manage only partial, modular tasks within larger systems."

A physicist at NASA:
"Accidents are an inevitable part of any technological development. While unfortunate, they offer critical opportunities to learn and improve safety moving forward."

Georges Khneysser, Founder QBIT Capital:
"Given the potential impact of AI on humanity, it may be wise to implement a 'kill switch,' a mechanism to shut down AI systems entirely if necessary."

ESSAY 11

AI AND INTELLECTUAL PROPERTY: EVOLUTION, DISRUPTION, OR BOTH?

By Markus Hauschild, Pasadena

Recently, and in response to President Trump's Executive Order "Removing Barriers to American Leadership in Artificial Intelligence" [1], the United States Patent and Trademark Office issued a new guidance [2] on inventorship for AI-assisted inventions to clarify the implications of using AI in the process of making inventions. Or, in other words, **who can be an inventor?** Briefly summarized, the guidance states that only natural persons can be properly named as inventors on patent applications - a hallmark of U.S. patent law that remains valid even as AI increasingly contributes to inventions. The guidance explicitly clarifies that "like any tool, while AI systems may assist inventors, such tools do not qualify for or elevate such assistance to inventor status." In conclusion, this guidance clarifies and normalizes the use of AI tools as part of the inventive process. To an inventor, this may feel like a natural progression or evolution rather than a disruption.

For those who rely on another form of intellectual property protection, namely, copyright, the situation is quite different. Copyright exists to protect the creative works of humans. However, arguably, the more recent AI applications can generate works that might qualify for copyright protection, had a human created them. This inevitably raises some questions:

Can AI-generated works be copyrighted? While the full answer to this question is complex, the short and simplified answer is surprisingly similar to the answer to the patentability question: AI may be used as a tool. In this case, copyright protection may apply to the human-created aspects of the creative work. In contrast, there is no copyrightability for works generated solely by AI. It seems reasonable to expect that what exactly is copyrightable with the use of AI tools will evolve over time. Photographs may be a good analogy. While the copyrightability of photographs is broadly accepted, this was not always the case. Initially, photography was interpreted as limited to reproducing scenery, rendering it ineligible for copyright protection. Over time, however, artistic aspects such as the creative choices involved in photography were recognized, and nowadays photographs are firmly copyrightable. We may eventually witness a similar historical journey that reflects a changing view of works generated with the assistance of AI. This has been true for various forms of expression, as the U.S. Copyright Office kept expanding protection by copyright law throughout history, which was initially limited to books, charts, and maps, to include categories such as musical compositions, visual artworks, motion pictures, sound recordings, computer programs, and architecture.

One may also wonder about **whether and how AI can cause copyright infringement**. Phrased differently, is the use of copyrighted materials for AI training covered under the "fair-use doctrine" (which balances creator rights with public access by permitting certain limited use of copyrighted material without permission for purposes such as criticism, teaching, news, research, etc.)? Or does it constitute infringement? A content creator may care about this question, for example, because a resulting AI-generated work may compete with the content creator's copyrighted work, thereby posing a direct threat to the content creator's commercial success. The answer to this question is that we presently have no definitive answer. Lawsuits have been filed, and their outcomes could redefine how to train AI models and whether AI companies must license materials from content creators. Should the courts rule against the AI companies, they may need to seek licensing agreements with content creators, publishers, and other content owners, which could significantly impact the business models of AI companies. Even the U.S. Copyright Office acknowledges the significant degree of uncertainty, stating that "it is not possible to prejudge litigation outcomes" and that "some uses of copyrighted works for generative AI training will qualify as fair use, and some will not" [3].

While the legal landscape remains dynamic, **content creators are understandably worried.** AI-created content will likely compete with human-created content for the content consumers' attention, suggesting that they will dedicate a smaller share of their attention to human-created content. Interestingly, though, a recent article in *The Economist* [4] suggests that the opposite may be true. The author discusses the effects of the spreading of slop across online platforms on content creators. Slop, the sometimes low-quality, often generic, or inaccurate content AI produces, affects different groups of content creators in different ways. On the one hand, the author concludes that the democratization of content creation by AI provides a considerable number of content creators with a living and turbocharges ordinary people's ability to produce and monetize AI-generated content. The article also disproves the notion that this has detracted attention from the biggest stars. Instead, because content consumers rely on algorithms to point them to the best content, many end up in the same place. On the other hand, the article points out that those content creators who are not successful enough to reach the top will be forced to share the scraps with millions more rivals than before.

Thus, we may safely conclude that **AI will produce both winners and losers among content creators**. Nevertheless, with AI and its adoption proceeding at a seemingly ever-increasing pace, it seems far more difficult to accurately predict the winners and losers. What appears to be clearer is this: Content creators and innovators will be affected by the rapid ascendance of AI in some form. Resistance is futile.

* * *

Markus Hauschild is a patent engineer at Osha Bergman Watanabe & Burton LLP, an international intellectual property law firm that focuses on innovations in robotics, artificial intelligence and other technologies. His background includes graduate and postdoctoral research in biomedical engineering and neuroscience at the University of Southern California and Caltech, and robotics and control at the German Aerospace Center (DLR). He also pursues a second career as a mountaineer and has summited major peaks on various continents, including Mt. Everest.

REFERENCES

[1] Exec. Order No. 14179, 90 FR 8741. (2025). Retrieved from https://www.govinfo.gov/content/pkg/FR-2025-01-31/pdf/2025-02172.pdf
[2] Revised Inventorship Guidance for AI-Assisted Inventions, 90 FR 54636. (2025). Retrieved from https://www.govinfo.gov/content/pkg/FR-2025-11-28/pdf/2025-21457.pdf
[3] Copyright and Artificial Intelligence, Part 3: Generative AI Training. (2025). Retrieved from https://www.copyright.gov/ai/Copyright-and-Artificial-Intelligence-Part-3-Generative-AI-Training-Report-Pre-Publication-Version.pdf
[4] The Economist. (2025, October 16). Sloponomics: who wins and loses in the AI-content flood? *The Economist*. Retrieved from https://www.economist.com/business/2025/10/16/sloponomics-who-wins-and-loses-in-the-ai-content-flood

WHO OWNS CREATIVE WORK?

Pamela Z, Composer and Media Artist:
"AI has made ownership of creative work confusing. If a machine generates the content, who owns it? The prompter? The software company? As artists, we're losing clarity and control over what we create."

Roland Emmerich, Film Producer, Director, and Screenwriter:
"The problem with AI is that it draws from everything, from the entire internet. Legally, you can't really do that. Most people don't even know where something comes from. That's totally frightening. But you can't stop technology. That's absolutely impossible."

Georges Khneysser, Founder QBIT Capital:
"AI doesn't 'steal' ideas; it learns from existing works in much the same way that young artists learn, by studying galleries, engaging with mentors, and exploring creative communities."

Robbert Baruch, Senior VP Public Affairs, Universal Music Group:
"Copyright protection must apply only to the work of real artists and composers. Purely AI-generated output should not qualify."

CNN News, 11 June 2025:
"Disney and Universal team up to sue AI photo generator Midjourney, claiming copyright infringement."

REGULATING THE WILD WEST

A Physicist at NASA:
"Cultural differences between Europe and the U.S. are reflected in their regulatory approaches: Europe tends to regulate first and develop later, while the U.S. often takes the opposite route – developing first, then regulating in response. Especially in California, the tech sector operates in what some describe as a regulatory 'Wild West.'"

Rear Admiral Grace Hopper, Computer Science Pioneer (1906–1992):
"It's better to ask for forgiveness than to ask for permission."

Markus Hauschild, Patent Engineer and Mount Everest Summiteer:
"Right now, we're in the 'Wild West' phase of AI: growth is exponential, no one knows what the next breakthrough will be, and regulation is struggling to keep up. In the long term, I'm optimistic as long as AI develops organically rather than explosively. Independent oversight will likely be essential."

Stefan Hartung, CEO Bosch GmbH:
"Europe is unnecessarily delaying its AI future with excessive regulation. [...] we will regulate ourselves to death, because we are trying to regulate against technological progress."

WHO SETS THE RULES – WHO PAYS THE PRICE?

A physicist at NASA:
"Robust, international regulation is essential, and must come from independent authorities, not from the companies themselves. It is not the role of private industry to set ethical or safety standards, but rather to comply with them. Companies act in pursuit of profit and will not necessarily prioritize ethics or social responsibility unless compelled by law."

Martin Moszkowicz, Film Producer, Constantin Film:
"Regulation is necessary to avoid potential dangers and prevent misuse. I'm not worried about AI 'taking over the world.' I consider that idea nonsense."

Richard Tuttobene, Film and TV Composer, Pianist, and Recording Artist:
"I'm drawn to music that moves forward, explores new ideas, and carries new messages. What frustrates me is the stagnation I see today: Studio execs and streaming services shying away from risk in favor of guaranteed returns, avoiding bold decisions that have a chance of succeeding but with greater impact. My greatest concern is AI in the hands of unrestrained capitalism."

ESSAY 12

WHEN ALGORITHMS MEET ACCOUNTABILITY: AI AND THE FUTURE OF JOURNALISM

By Lukas Görög, Zurich

In the spring of 2023, I sat in a newsroom in Tbilisi, Georgia, training journalists on how to detect AI-manipulated images. A few months earlier, a deepfake video of a Georgian opposition leader had spread across social media, nearly destabilizing an already fragile political situation. The journalists I worked with were not technophobes; they were curious, resourceful, and determined, yet also overwhelmed. The tools used to create the fake had been freely available online. The tools to detect it? Those required expertise, resources, and time that most newsrooms simply did not have.

This moment crystallized something I had been observing for years: Artificial intelligence is simultaneously the greatest threat to journalism and its most promising ally. The question is not whether AI will transform journalism; it already has. The question is whether journalists will shape this transformation or be shaped by it.

THE PROMISE: AUGMENTING HUMAN INTELLIGENCE

Let me be clear from the outset: I am not a technopessimist. In my work at the Neue Zürcher Zeitung (NZZ), one of the oldest and most respected newspapers in the German-speaking world, I have seen firsthand how AI can elevate journalism rather than diminish it.

Consider the daily reality of a modern newsroom. Journalists are drowning in data: press releases, social media posts, government documents, leaked databases. The Panama Papers, for instance, comprised 11.5 million documents. No human team could have analyzed them without computational assistance [1]. AI-powered tools can now sift through vast datasets, identify patterns, flag anomalies, and suggest leads that would take human researchers months to uncover.

At the NZZ, we have developed internal AI tools that assist journalists in multiple ways: from automated transcription and translation to intelligent content recommendations that help reporters discover relevant archival material. But equally important has been learning to use external AI platforms efficiently, integrating third-party tools such as ChatGPT, Claude, and specialized verification services into our workflows. We have built systems that analyze reader engagement patterns, that do not chase clicks but understand which stories resonate and why. These tools, whether built in-house or adopted from outside, do not replace journalistic judgment; they inform it.

The potential extends further. AI can democratize access to investigative journalism by reducing the resource barriers that have traditionally limited such work to large, well-funded outlets. A small regional newspaper in Austria, equipped with the right AI tools, can now undertake the kind of data journalism that was once the exclusive domain of *The New York Times* or *The Guardian*.

THE PERIL: A CREDIBILITY CRISIS IN THE MAKING

Yet for every opportunity AI creates, it introduces a corresponding risk. And in journalism, the stakes could not be higher.

Research has demonstrated that deepfakes contribute to what scholars call "generalized indeterminacy and cynicism," in which people become uncertain about whether *any* media content can be trusted [2]. A landmark study by Vaccari and Chadwick found that exposure to deepfakes makes people more likely to feel uncertain than to be actively deceived. But this uncertainty, in turn, significantly reduces trust in news on social media. The implications are profound: We may be heading toward an epistemic crisis where the very concept of visual evidence loses its evidentiary value.

I have witnessed this firsthand. In my consulting work with media organizations in Myanmar, the Balkans, and Mexico, I have seen how AI-generated content can be weaponized to inflame ethnic tensions, discredit political opponents, and undermine public trust in institutions. The tools are cheap, accessible, and improving rapidly. The defenses? They lag behind.

There is also the "model collapse" problem, as researchers call it. A 2024 study published in *Nature* demonstrated that, when AI models are trained on content generated by previous AI models rather than human-created data, they experience progressive degradation, eventually producing outputs that drift far from reality [3]. As synthetic content floods the internet, it creates a dangerous feedback loop. If the raw material of AI training becomes contaminated with misinformation, bias, and synthetic noise, the systems journalists rely on will inherit these flaws. The study's authors warn that this "degenerative learning process" could fundamentally compromise the reliability of future AI systems.

THE PARADOX: AI AS BOTH WEAPON AND SHIELD

Here is the paradox at the heart of AI and journalism: The same technologies that enable disinformation can also be used to combat it.

At the NZZ, we have invested significantly in AI-powered verification tools. These systems can detect image manipulation, identify inconsistencies in metadata, cross-reference claims against trusted databases, and flag potential deepfakes for human review. But technology is only part of the solution: We have also expanded a dedicated OSINT (Open Source Intelligence) team that combines human expertise with AI capabilities to investigate disinformation at scale. We are not alone. Organizations like the Reuters Institute, Full Fact, and the Duke Reporters' Lab are developing AI-driven fact-checking systems that can operate at scale [4].

But technology alone is not sufficient. In my experience training journalists across 12 countries, I have learned that the most effective defense against AI-enabled disinformation is not another algorithm; it is a well-trained journalist who understands how these systems work. AI success in newsrooms requires both education and infrastructure working in tandem [5]. That is why I founded the Academy for Artificial Intelligence in Zurich and Vienna: to ensure that media professionals are not merely consumers of AI tools but informed, critical users who can interrogate their outputs and understand their limitations.

The fourth estate has a unique responsibility in democratic societies. AI can streamline operations so that journalists can focus on the parts of the job that only humans can do. But this works only if journalists possess the knowledge to use these tools wisely and to recognize when they are being misused.

THE HUMAN ELEMENT: WHAT AI CANNOT REPLACE

For all its capabilities, AI lacks something fundamental: It cannot care. It cannot feel the moral weight of exposing corruption. It cannot build trust with a source over months of careful relationship-building. It cannot sit across from a whistleblower and recognize the fear in their eyes.

At its best, journalism is a profoundly human endeavor that requires empathy, ethical judgment, and the courage to speak truth to power. These are not capabilities that can be encoded in a neural network.

Research confirms what intuition suggests: Readers trust news more when they know humans are involved in its creation. A study from the University of Kansas found that, when readers believe AI contributed to news production, their trust in its credibility decreases, even when the content quality is objectively comparable to human-written articles [6]. This finding underscores a deeper truth: Trust in journalism is not determined solely by accuracy or style but by perceptions of authorship and accountability.

I often tell the journalists I train: AI will not take your job. But a journalist who understands AI might take the job of one who does not. This is not meant as a threat; it is a call to action. The future belongs to hybrid intelligence: human creativity and judgment augmented by machine speed and scale.

There is also a market dimension to this. In a world flooded with synthetic content, authentic human journalism may become more valuable precisely because of its scarcity. As generative AI drives down the costs of producing inauthentic content, the value of human authenticity and editorial oversight is being reaffirmed rather than diminished [7]. Readers may increasingly seek out sources they trust. And trust, ultimately, is built by humans, not machines.

THE PATH FORWARD: PRINCIPLES FOR AN AI-AUGMENTED PRESS

What, then, should journalism do? Based on my years of work at the intersection of AI and media, I propose four principles:

- *First, invest in AI literacy.* Every journalist should understand the basics of how large language models (LLMs) work, how images can be manipulated, and how to verify digital content. This is no longer a specialized skill – it is foundational.
- *Second, develop a clear AI roadmap.* News organizations need strategic clarity about where AI fits into their operations. Without a roadmap, newsrooms risk either missing opportunities or adopting tools haphazardly without understanding their implications [8]. A roadmap forces organizations to articulate their values and align technology adoption with editorial mission.
- *Third, collaborate across borders.* Disinformation does not respect national boundaries, and neither should our defenses. International coalitions of journalists, technologists, and researchers are essential for developing robust verification tools and sharing best practices.
- *Fourth, remember what makes journalism matter.* In the rush to adopt new technologies, we must not lose sight of the values that define quality journalism: accuracy, fairness, independence, and accountability. AI should serve these values, not supplant them.

CONCLUSION: MEETING AI EYE TO EYE

The title of this book – *AI to Eye* – captures something essential: To engage with artificial intelligence responsibly, we must look at it directly, unflinchingly, and with clear eyes. We must see both its potential and its dangers.

For journalism, the stakes are existential. A free press is the immune system of democracy. If that immune system is corrupted by misinformation, overwhelmed by synthetic content, or atrophied by complacency, the consequences extend far beyond the media industry.

But I remain optimistic. In newsrooms from Zurich to Tbilisi, from Vienna to Mexico City, I have met journalists who are rising to this challenge, learning new tools, asking hard questions, and refusing to cede their essential role to machines. They understand that AI is neither savior nor destroyer. It is a mirror, reflecting our intentions back at us.

The question is not what AI will do to journalism. The question is what journalists, and we as a society, will do with AI, and what AI will do to us. The answer is still ours to write.

* * *

Lukas Görög worked as the AI and Data Strategy Lead at the Neue Zürcher Zeitung (NZZ) in Zurich and is the founder of the Academy for Artificial Intelligence in Vienna and Zurich. He educates and consults various newsrooms and media organizations in the areas of AI and data. Educated at the TU Wien, TU Bratislava, and LSE in London, he supports media organizations across Europe, North America, and Asia in AI integration, misinformation detection, and journalist training. He has conducted over 400 AI workshops for media professionals worldwide.

REFERENCES

[1] Obermaier, F., & Obermayer, B. (2016). *The Panama Papers: Breaking the story of how the rich and powerful hide their money*. Oneworld Publications.
[2] Vaccari, C., & Chadwick, A. (2020). Deepfakes and disinformation: Exploring the impact of synthetic political video on deception, uncertainty, and trust in news. *Social Media + Society, 6*(1), 1–13. https://doi.org/10.1177/2056305120903408
[3] Shumailov, I., Shumaylov, Z., Zhao, Y., Papernot, N., Anderson, R., & Gal, Y. (2024). AI models collapse when trained on recursively generated data. *Nature, 631*(8022), 755–759. https://doi.org/10.1038/s41586-024-07566-y
[4] Graves, L., & Cherubini, F. (2016). *The rise of fact-checking sites in Europe*. Reuters Institute for the Study of Journalism.
[5] Görög, L. (2023, September). *AI success requires education and infrastructure*. INMA Blog. Retrieved from https://www.inma.org/blogs/GenAI/post.cfm/ai-success-for-publishers-requires-education-infrastructure-as-well-as-tech
[6] Appelman, A., Bien-Aimé, S., Jia, H., & Wu, M. (2024). Reader perceptions of AI involvement in news production. *Communication Reports, 38*(1), 46–58. https://doi.org/10.1080/08934215.2024.2424553

[7] MEHTA, I. (2023). *Authenticity in the AI content era will not come cheap*. Markkula Center for Applied Ethics, Santa Clara University.
[8] GÖRÖG, L. (2025, February). *News companies need a clear AI road map for success*. INMA Blog. Retrieved from https://www.inma.org/blogs/GenAI/post.cfm/news-companies-need-a-clear-ai-road-map-for-success

AI AS A FORCE MULTIPLIER IN JOURNALISM

Helge Fuhst, Journalist and TV Moderator, Head of ARD "Tagesthemen":
"AI opens up new possibilities in journalism, streamlining workflows and enabling tools like advanced translation to help national reporting reach global audiences."

Melanie Amann, Journalist, Deputy Editor-in-Chief of "Der Spiegel":
"Thanks to AI, it is going to be easier for us to work on big investigative projects. […] We can standardize and minimize the production costs on the standard news business, for example, […] like the bread-and-butter stuff. […] We can really facilitate things so that our journalists can focus on the stuff that we feel only they can do, only human beings can contribute to the news […] I think we should embrace AI in that context."

Steven Walter, Curator, Artistic Director of the Beethovenfest Bonn:
"In my curatorial work, AI can greatly assist with logistics such as research, correspondence, program curation, meeting notes, budgeting. Tasks that once required full teams can now be done by individuals. But this raises concerns: are we losing valuable discursive space and team-based reflection in the process?"

WHO GETS SEEN AND RECOGNIZED?

Melanie Amann, Journalist, Deputy Editor-in-Chief of "Der Spiegel":
"I am very skeptical that our coverage and our high-quality journalism will still reach people. I mean, we see the Gemini summaries that Google offers for their searches [...]. How do people still find their way to our websites if they search for something and find the answer right away? Who needs the story that contains the answer? This means a huge loss of traffic to our websites and the danger that people will simply not consume our journalism anymore, but just a very digested and condensed summary of it. Combined with the lack of reading serious news sources, this will contribute to polarization. I have no doubt."

Marty Kaplan, Professor of Communication and Journalism, University of Southern California:
"AI-generated content, whether music, film, or visual art, is becoming nearly indistinguishable from human-made work, enabling business models that avoid licensing fees, for example, by using AI-generated music in films. In this context, human-made art may only retain value if originality can be preserved and clearly recognized by audiences."

WHEN TRUST BECOMES THE SCARCE RESOURCE

Helge Fuhst, Journalist and TV Moderator, Head of ARD "Tagesthemen":
"In today's fragmented media landscape, serious journalism is drowning in a flood of misinformation, undermining trust, quality, and societal stability."

Ranga Yogeshwar, Science Journalist and TV Presenter:
"We are blurred with artificial pictures and voices. At the end, we will have a credibility crisis, because everything you see in the media can be fake, nobody knows who to trust. Therefore, I believe the non-use of AI can be a unique selling proposition."

Kathryn Starkey, Professor of German Studies, Stanford University:
"I worry about students in the US: Social Media, the Internet and AI are challenging to navigate. Students often don't know where to go to get the truth! Moreover, truth is often perceived as a subjective category."

PISSING IN THE SOUP

Pamela Z, Composer and Media Artist:
"What worries me is how this thing learns: it absorbs everything it's fed, then creates new content that feeds back into the system. It's like pissing in the soup, a garbage-in, garbage-out problem."

Ranga Yogeshwar, Science Journalist and TV Presenter:
"I believe that AI is going to induce a self-intoxication of available information. Within the next ten years, much of the content on the internet may become synthetic or artificially generated. The main problem will be to find credible information. We're already seeing a major issue with LLMs: They often fail to provide trustworthy information, largely because you can't create a closed-loop system where AI is trained on content previously generated by other AI, detached from real-world grounding. The applications are perhaps going to kill themselves."

Kathryn Starkey, Professor of German Studies, Stanford University:
"I don't think writers or poets will go extinct. Generative AI cannot generate anything original. ChatGPT and other platforms can only reproduce what is input. As more texts are produced via AI and fed back into these models, the input will become less and less original. In a society in which many texts are created using AI, writers producing original work will become more valuable."

ESSAY 13

COULD ONE STEER HUMANS AND SOCIETIES WITH GENERATIVE AI?

By Dirk Helbing, Zurich

We have all heard the slogans: "Big Data is the new oil!" and "Artificial Intelligence is the new motor." But what does that really mean? Will we live in data-driven societies? Will AI make our decisions? Will AI be "our final invention" [1], given that AI is now coding software and also increasingly doing what humans have been doing in the past, fueled by the knowledge previously created by humanity? Where are we heading?

As it turns out, while most people think we are living in the information society, others have announced that we are already entering a new historical era [2]. Having gone through the hunter-gatherer society (Society 1.0), the agrarian society (Society 2.0), the industrial society (Society 3.0), and the information society (Society 4.0), "Society 5.0" is next and will result from the "fourth industrial revolution" (or "industry 4.0") - so we are told [3]. The latter is characterized by technologies such as AI, advanced robotics, and gene-editing, one can read. So, what is in store for us?

FROM THE INTERNET OF THINGS TO SMART CITIES

First of all, we will increasingly see cyberphysical systems, which means that the virtual and real worlds will grow together. One result is the "Internet of

Things" ("IoT") – a myriad of sensors connected to the Internet to measure environmental data, while wireless actuators will trigger changes.

Smart factories and smart cities are interesting areas of application. Imagine, for example, a traffic-control center that steers traffic flows based on real-time measurements, using sophisticated algorithms that enable automated control. Such applications will increasingly spread to other areas of society as well. Is the automation of society next [4]? And, if so, how far will it go? One contribution to the "Smart City Charta," for example, discusses a hypothetical "postdecision, postvoting society" [5]: "Because we know exactly what people do and want, there is no need to vote. Behavior-related data can replace democracy as a societal feedback system."

This would be a data-driven society, in which cybernetics would eventually replace governance. Accordingly, information feedback would increasingly replace traditional decision-making of politicians and people. If we just had good enough Big Data and powerful enough AI to see the patterns in it, to make sense of it, and take proper action, visionaries argue, this could be more efficient than the conventional governance approach – and it could also lead to better results than human decision-making, given our cognitive limits.

WILL DATA-DRIVEN SOCIETIES BRING US PARADISE?

Will "surveillance capitalism" [6], which has become surprisingly good at inferring real-world activities and even our thoughts, emotions, and behaviors, benefit us here? Could data-driven algorithms guess (or shape) our desires and eventually take steps to satisfy them? Could AI systems, one day, by continuously observing and permanently listening to us, make happen what we wish for? These would have to be generative AI systems and robotic tools, of course. Imagine one would feed them not only with texts, voice control, pictures, and movies, but also with data from the Internet of Things. Imagine they could also soon produce architectural designs, plan our cities, manage our finances, organize our lives, and perform much of our current work. Would we then live in "paradise," with our wishes effortlessly coming true?

Well, there are at least two caveats. First, this would probably require the capability to guess what's on our minds. Second, it may require the system to change other people's thoughts (to make them agree with the materialization of your own wishes). In other words, for the wishes of other people to come true, your thoughts would have to be manipulated, too. The manipulation of minds would become part of that envisioned "paradise."

While this sounds like science fiction, such a system may someday be actually built, using new technologies called Brain Computer Interfaces (BCIs). However, since most people would probably not want to undergo a surgery to have a brainchip implanted, the question arises whether there are also less invasive ways of creating BCIs. In fact, nano-neurotechnology may be a feasible approach. But how would it work?

"THE INTERNET OF BODIES IS HERE"

This brings us to the "Internet of Bodies." In 2020, an announcement of the World Economic Forum (WEF) entitled "Tracking How Our Bodies Work Could Change Our Lives" surprised the world [7]: "We're entering the era of the 'Internet of Bodies': collecting our physical data via a range of devices that can be implanted, swallowed, or worn." Only a few months later, the WEF announced [8]: "The Internet of Bodies Is Here" and "Recent technological advancements have ushered in a new era of the 'internet of bodies' (IoB), with an unprecedented number of connected devices and sensors being affixed to or even implanted and ingested into the human body."

What are they talking about? Scientifically speaking, the "Internet of Bodies" is called the "Internet of [Bio-]Nano Things" ("Io[B]NT") [9]. "Converging technologies" can now connect information and communication technologies with [bio- and] nanotechnologies [10]. Such nanotechnologies are based on nanoparticles – particles of up to 100 nanometers in size, that is, less than one ten-thousandth of a millimeter, invisible to the human eye, fitting into human cells. When absorbed via food, drinks, medicine, or the environment, such nanoparticles can be used in various ways.

With magnetic nanoparticles, carbon-based nanomaterials, or quantum dots, for example, it becomes possible to read out data on the functioning of our bodies, using energy beams (suitable sound or electromagnetic radiation [11]). Notably, this method requires no "nanochips" or "smart nanoparticles," as the data can be read out using physics principles (such as absorption, reflection, or stimulated photon emission). Think of how an injected contrast medium works during an MRI scan of the body. In the meantime, real-time *in vivo* biosensing is an active area of research [12]. One day, such approaches may allow one to monitor health data outside clinical settings in everyday situations, without visiting a doctor.

BRAIN ACTIVITY MAPPING

Already back in 2013, Barack Obama announced a major research project [13]: "The BRAIN Initiative® aims to deepen understanding of the inner workings of the human mind and to improve how we treat, prevent, and cure disorders of the brain," the related webpage said. More details can be found in a specialist publication [14]: "Nanoscience and nanotechnology are poised to provide a rich toolkit of novel methods to explore brain function by enabling simultaneous measurement and manipulation of activity of thousands or even millions of neurons. We and others refer to this goal as the Brain Activity Mapping Project."

Is the aim of this to target human brains? There has been quite intensive research on technologies to read out brain data. One particularly interesting method is known as "optogenetics," which explores ways to stimulate nerve cells, genetically modified or not, by using light pulses [15]. In 2021, the British Ministry of Defense stated [16]: "The future of human augmentation should not ... be decided by ethicists or public opinion..." and "It could be argued that treatments involving novel vaccination processes and gene and cell therapies are examples of human augmentation already in the pipeline." This was announced after the US had already been working on injectable neurotechnologies, also known as "noninvasive neurotechnologies," which do not need a brain surgery to work [17].

The goal of such efforts is not just to *infer* human thoughts, as surveillance capitalism has been able to do for some time to a surprising extent, but also to make minds *readable*. And where data can be read *out*, it becomes possible to write data *in*. While such approaches are still experimental, they are expected to eventually allow one to induce specific thoughts in our own minds and even transmit thoughts to other brains using specialized BCIs.

The military was interested in such "synthetic telepathy" already 20 years ago [18]. In 2014, it was reported to be within reach [19]. Given that powerful AI is good at finding patterns in Big Data, would AI also be able to reveal human thought patterns? When would we see the first applications?

In 2024, the Silicon Valley visionary Ray Kurzweil voiced his expectation that we *would* merge with AI, saying [20]: "We're going to be a combination of our natural intelligence and our cybernetic intelligence and it's all going to be rolled into one. Making it possible will be brain-computer interfaces which ultimately will be nanobots – robots the size of molecules – that will go noninvasively into our brains through the capillaries." In the same year, the UNESCO elaborated a recommendation on the ethics of neurotechnology for the United Nations [21].

NEUROCAPITALISM AHEAD?

Considering all these stunning technological developments, could specialized forms of generative AI even learn to remotely steer human behavior and societies? If yes, how many parameters would one need to control human behavior, how many to control an entire society? How well would that work? And what political and economic systems could one build with this? "Neurocapitalism," for example [22,23]?

In contrast to "surveillance capitalism," "neurocapitalism" would not only *infer* our actions, thoughts, and feelings through mass surveillance – it would even try to technologically *shape* them. Eventually, one could weave neurotechnology (and capitalism) into every thought. Experiments suggest that it would be possible to hijack our dreams for advertising [24]. And already back in 2019, there were warnings about an upcoming "era of neurocapitalism": "Brain-reading tech is coming," a news article said [25], and "The law is not ready to protect us." This needs to be changed!

HUMAN RIGHTS AND HUMAN DIGNITY IN DANGER

Before we discuss the societal implications, let us summarize some preliminary conclusions:

- Global corporations have turned us into data mines for exploitation.
- Large-scale manipulation through personalized information has been happening already for some time (even before the emergence of possibilities for neuronal manipulation), e.g., via microtargeting, persuasive computing [26] and generative AI [27].
- WiFi, 5G, 6G, LEDs are not only communication networks or sources of light, but can also be used as parts of measurement (and control) networks.
- Not only can smart watches and medical devices be used to read out body data, but nanoparticles can be used as "wearables" as well, which makes us vulnerable!
- Human augmentation has military applications, and cognitive warfare is already a major concern [28].

I do not think businesses or states should be allowed to experiment and interfere with our bodies and brains without our invitation. We need laws and technological solutions to protect us! Steering people and societies in data-driven ways has become a real possibility. Even long before neurotechnology is applied at scale, we should ask ourselves: Are we at risk of heading towards a data dictatorship that increasingly takes the power over our lives away from us? In view of this dystopian potential, I plead for a new soc(iet)al contract that ensures that the use of technologies is individually benevolent, respectful, fair, and safe, driven not by profit or power.

I do not believe it is reasonable or legitimate for companies or states to take control over our personal, brain, and health data – or to control our thoughts, bodies, and actions remotely. The justification often given for such aims, namely, that the world can be optimized to perfection in this way and thereby turned into a paradise, does not hold. If that were the case, we would already have noticed it. But the dream has not materialized, for example, due to hallucinations and other problems produced by "black box algorithms" [29] and generative AI. Instead, we have slipped into a "polycrisis," where democracy and human rights, peace and planetary health appear to be in acute danger (considering also the adverse health effects of many nanoparticles).

There is an urgent need for a new approach to technology based on the principles of responsible innovation, which includes the precautionary principle. The current "social engineering" approach, based often on the idea of a "benevolent dictatorship," should be replaced by assisted self-organization and symbiotic coevolution. Value-sensitive design [30] can ensure that technology strengthens constitutional values and serves the people – not the other way around. This allows democracies to be upgraded by design [31]. For example, it should be easy for us to determine what is being done with our personal data and digital twins. That is what the right to informational self-determination implies.

We need enforceable rules that are compatible with human dignity and universal human rights. Instead of a supposedly optimal future, there should be good alternatives to choose from. Instead of social control, we need serendipity. Artificial intelligence should create opportunities for all and help us to improve the world together. Then, it would become *our* world again, not just an increasingly automated, technical artifact.

* * *

Dirk Helbing is Professor of Computational Social Science at the Department of Humanities, Social and Political Sciences. He is also associated with ETH Zurich's Department of Computer Science. His current studies discuss the digital society, smart cities as well as the challenges and opportunities of digital democracies. Among others, he coordinated the doctoral program "Engineering Social Technologies for a Responsible Digital Future" at TU Delft. He is also known for having led the FuturICT initiative. Furthermore, he is an elected member of the German Academy of Sciences "Leopoldina" and a member of the External Faculty of the Complexity Science Hub Vienna.

REFERENCES (weblinks accessed on 20 February 2026)

[1] Barrat, J. (2013). *Our final invention: Artificial intelligence and the end of the human era*. Thomas Dunne Books.
[2] Deguchi, A. et al. (2020). What is Society 5.0? In H. Deguchi et al. (Eds.), *Society 5.0: A people-centric super-smart society*. Springer.
[3] Wikipedia contributors. (n.d.). *Fourth industrial revolution*. Wikipedia. Retrieved from https://en.wikipedia.org/wiki/Fourth_Industrial_Revolution
[4] Helbing, D. (2026). How emerging nano-, neuro- and quantum technologies enable emerging cybernetic societies. ResearchGate preprint. https://www.researchgate.net/publication/399951376
[5] Bundesministerium für Umwelt, Naturschutz, Bau und Reaktorsicherheit. (2017). *Smart City Charta (Langfassung)* (p. 43). Retrieved from https://www.scribd.com/document/615445985/smart-city-charta-langfassung
[6] Zuboff, S. (2019). *The age of surveillance capitalism*. PublicAffairs.
[7] World Economic Forum. (2020, June 4). *Urban transformation: Tracking how our bodies work could change our lives*. Retrieved from https://www.weforum.org/stories/2020/06/internet-of-bodies-covid19-recovery-governance-health-data/
[8] World Economic Forum. (2020, August 6). *The internet of bodies is here: Tackling new challenges of technology governance*. Retrieved from https://www.weforum.org/publications/the-internet-of-bodies-is-here-tackling-new-challenges-of-technology-governance/
[9] Al-Turjman, F. (2019). *Internet of nano-things and wireless body area networks (WBAN)*. Taylor & Francis.
[10] Helbing, D., & Ienca, M. (2024). Why converging technologies need converging international regulation. *Ethics and Information Technology, 26*, 15. https://doi.org/10.1007/s10676-024-09756-8.
[11] Celik, A., Salama, K. N., & Eltawil, A. M. (2022). The internet of bodies: A systematic survey on propagation characterization and channel modeling. *IEEE Internet of Things Journal, 9*(1), 321–345. https://doi.org/10.1109/JIOT.2021.3098028
[12] Eckert, M. A., & Zhao, W. (2013). Opening windows on new biology and disease mechanisms: Development of real-time in vivo sensors. *Interface Focus, 3*(3), Article 20130014. https://doi.org/10.1098/rsfs.2013.0014
[13] The BRAIN Initiative. (n.d.). *Mission*. Retrieved from https://www.braininitiative.org/mission/
[14] Alivisatos, A. P. et al. (2013). Nanotools for neuroscience and brain activity mapping. *ACS Nano, 7*(3), 1850–1866. https://pubs.acs.org/doi/10.1021/nn4012847
[15] Delbeke, J. et al. (2017). And then there was light: Perspectives of optogenetics for deep brain stimulation and neuromodulation. Frontiers in Neuroscience 11:633. https://doi.org/10.3389/fnins.2017.00663
[16] UK Ministry of Defence. (2021). *Human augmentation: The dawn of a new paradigm*. Retrieved from https://assets.publishing.service.gov.uk/government/uploads/system/uploads/attachment_data/file/986301/Human_Augmentation_SIP_access2.pdf
[17] Battelle. (2019, May 20). *Battelle-led team wins DARPA award to develop injectable, bi-directional brain-computer interface*. Retrieved from https://www.battelle.org/insights/newsroom/press-release-details/battelle-led-team-wins-darpa-award-to-develop-injectable-bi-directional-brain-computer-interface

[18] Bland, E. (2008, October 13). Army developing "synthetic telepathy." *NBC News*. Retrieved from https://www.nbcnews.com/id/wbna27162401
[19] Iozzio, C. (2014, October 2). Scientists prove that telepathic communication is within reach. *Smithsonian Magazine*. Retrieved from https://www.smithsonianmag.com/innovation/scientists-prove-that-telepathic-communication-is-within-reach-180952868/
[20] Corbyn, Z. (2024, June 29). AI scientist Ray Kurzweil: We are going to expand intelligence a millionfold by 2045. *The Guardian*. Retrieved from https://www.theguardian.com/technology/article/2024/jun/29/ray-kurzweil-google-ai-the-singularity-is-nearer
[21] UNESCO, Outcome document of the first meeting of the AHEG: first draft of a Recommendation on the Ethics of Neurotechnology (first version). Retrieved from https://unesdoc.unesco.org/ark:/48223/pf0000389768; UN General Assembly, Impact Impact, opportunities and challenges of neurotechnology with regard to the promotion and protection of all human rights. Retrieved from https://documents.un.org/doc/undoc/gen/g24/133/28/pdf/g2413328.pdf
[22] Pykett, J. (2013). Neurocapitalism and the new neuros: Using neuroeconomics, behavioural economics and picoeconomics for public policy. *Journal of Economic Geography, 13*(5), 845–869. https://doi.org/10.1093/jeg/lbs039
[23] Cassidy, F. (2021, October 19). *Neurocapitalism – The extraction and exploitation of our thoughts*. Medium. Retrieved from https://medium.com/@feargal.cassidy1/neurocapitalism-the-extraction-and-exploitation-of-our-thoughts-c8435652ddb8
[24] Delbert, C. (2021, July 8). Advertisers are hijacking your dreams, scientists say. *MIT Media Lab*. Retrieved from https://www.media.mit.edu/articles/advertisers-are-hijacking-your-dreams-scientists-say/
[25] Samuel, S. (2019, December 20). Brain-reading tech is coming. The law is not ready to protect us. *Vox*. Retrieved from https://www.vox.com/2019/8/30/20835137/facebook-zuckerberg-elon-musk-brain-mind-reading-neuroethics
[26] Kaiser, B. (2020). *Targeted*. HarperCollins UK.
[27] Lin, H. et al. (2025). Persuading voters using human-artificial intelligence dialogues. *Nature* 648, 394–401. Retrieved from https://www.nature.com/articles/s41586-025-09771-9. https://doi.org/10.1038/s41586-025-09771-9
[28] Pujol, I. (2024, September 2). *Cognitive war turns the mind into battleground*. IE University. Retrieved from https://www.ie.edu/insights/articles/cognitive-war-turns-the-mind-into-battleground/
[29] Pasquale, F. (2016). *The black box society: Secret algorithms that control money and information*. Harvard University Press.
[30] Friedman, B., & Hendry, D. G. (2019). *Value sensitive design: Shaping technology with moral imagination*. MIT Press.
[31] Helbing, D. et al. (2023). Democracy by design: Perspectives for digitally assisted, participatory upgrades of society. *Journal of Computational Science, 71*, Article 102061. https://doi.org/10.1016/j.jocs.2023.102061

LEARNING FROM PAST MISTAKES

Martin Keller, President of the Helmholtz Association:
"If I had a magic wand, I would wish for humanity to finally learn from past mistakes instead of repeating them every few years. In that spirit, AI must be applied fairly, so it serves everyone as a source of progress and benefit."

A Physicist at NASA:
"AI presents significant risks to society, some of which may become irreversible if not addressed early."

ADDICTION IS A SYSTEMIC RISK

Russel Wald, Executive Director, Institute of Human-Centered AI, Stanford University:
"I am especially concerned about aspects of addiction."

Georges Khneysser, Founder QBIT Capital:
"AI applications may be even more addictive than social media. While social media stimulates us with short bursts of dopamine, AI engages our desire for knowledge, making us feel smarter and more capable."

Haewon Jeong, Professor of Computer Engineering, UC Santa Barbara:
"Even if we solved the problem of addiction to AI-generated social media, new forms of addiction would likely emerge."

MIND CAPTURE

An Earth Scientist at NASA:
"AI-driven algorithms in social media can influence public behavior through subtle adjustments in the information people receive, often

without their conscious awareness. This influence can be used for good, encouraging beneficial choices without individuals realizing it."

Bruno Gandlgruber, Professor of Economics and Future Studies, UNESCO Chair, UAM Mexico City:
"Behavioral changes should build on a democratic process of opinion-making. Brainwashing people's mind via social media would not be democratic."

Russel Wald, Executive Director, Institute of Human-Centered AI, Stanford University:
"Whether AI will promote societal polarization remains to be seen. But I am skeptical. Where there's smoke, there's fire."

DEMOCRACY UNDER PRESSURE

Thomas Mann, Novelist, Nobel Prize Winner in Literature (1875–1955):
"... it's not that the elites notice it; What counts are the masses. Intelligence no longer matters at all: Whoever speaks out is lynched or 'hanged.' The question is: what mode of existence will intelligence take now, and since it no longer has any means of influence, how will it at least manage to preserve its critique for posterity?" (Lugano, 5 April 1933)

Refik Anadol, Media Artist and Designer:
"The rapid growth of AI is disrupting our economy and may deepen inequalities. The sooner we clarify what AI means for humanity, the better we can ensure fairness. Current AI tools are cheap and widely accessible, so that they provide a relatively democratic starting point. Yet the pace of development is absurdly fast, making it a major challenge to democratize access and application of AI worldwide."

Yoshua Bengio, AI-Pioneer:
"I started thinking about my grandchild, and I thought, in 20 years, he's going to be 22; will he have a life? Will he live in a democracy? Whoever controls very advanced AIs in the future will have huge power. And they could use that power in ways that are good for them, maybe, but not for most of us. Democracy is about sharing power. If the power is concentrated in the hands of a few, that is not democracy – that is a dictatorship."

Helge Fuhst, Journalist and TV Moderator, Head of ARD "Tagesthemen":
"The rise of AI and social media risks deepening polarization and weakening democracy, but also opens space for quality journalism to stand out."

POWER WITHOUT RESPONSIBILITY

Bruno Gandlgruber, Professor of Economics and Future Studies, UNESCO Chair, UAM Mexico City:
"There is a concentration of power within a small number of large companies. Ethics is not a big issue for those companies. But this is not an accusation rather than an economic fact. Society must guarantee to monitor the behavior of the companies, creating mechanisms of regulation and of accountability."

Emmanuel Mignot, Professor of Sleep Medicine, Stanford University & Recipient of the Breakthrough Prize:
"Leading people in the industry don't care about the future. They just enjoy the current ride."

A Software Engineering Manager at Google:
"People here aren't concerned about the future – they're focused on maximizing profit. They don't perceive any immediate problems, and if issues do arise, they'll just try to fix them as they come."

Martin Keller, President of the Helmholtz Association:
"I doubt pure capitalism can handle the disruption due to AI. We need a socially responsible, democratic market system, since companies and investors are too focused on profit alone."

Bruno Gandlgruber, Professor of Economics and Future Studies, UNESCO Chair, UAM Mexico City:
"Most companies only pretend to take care of ethical issues (for society, climate, etc.). This kind of 'green washing' can improve their market position; some companies even withhold relevant information that could weaken their market position."

CAN SOCIETY STEER THE OUTCOME?

Robert Habeck, Germany's Vice Chancellor and Federal Minister for Economic Affairs 2021–2025:
"Through the AI Act, the European Union has established a comprehensive legal framework for artificial intelligence that ensures its legally compliant and trustworthy deployment. To date, this is unique worldwide."

A Professor of Computer Science, University of Southern California:
"Will AI companies act responsibly, or be solely motivated by profit? Companies prioritize shareholder interests, so raising awareness of risks and options among shareholders and the public is essential. Informed shareholders can push for ethical practices, as they care about the future. However, when that future seems distant and immediate benefits are clear, encouraging responsible action is more difficult."

Renée Reizman, Interdisciplinary Artist and Writer:
"In a best-case scenario, AI generates enough wealth that people displaced from work can rely on functioning social services and a basic income, enjoying life without worry. However, our highly individualistic culture ('I worked hard for myself, no one helped me, so why should I help you') makes such a system unlikely without a major societal shift."

ESSAY 14

AFTER INTELLIGENCE: ON WHAT REMAINS HUMAN

By Robert Riener, Zurich and Los Angeles

Artificial intelligence (AI) has pushed the idea of intelligence itself into unfamiliar territory. As machines increasingly perform tasks once thought to require uniquely human intellect, the question is no longer simply *how* intelligent AI has become, but *what* remains of intelligence once performance alone is no longer decisive. This closing essay approaches intelligence from that vantage point – *after* the initial excitement, after the benchmarks, and after the headlines – to reflect on what intelligence signifies when humans and machines are placed side by side. It argues that comparison does not diminish human intelligence but instead reveals the deeper structures that distinguish it.

DEFINING INTELLIGENCE

A single definition has never easily pinned down human intelligence. Traditional psychological accounts describe it as the capacity for learning, reasoning, problem-solving, and adaptation. Encyclopedic sources, such as *Encyclopedia Britannica,* emphasize understanding, abstraction, and the ability to apply knowledge flexibly, while popular summaries, such as those found in Wikipedia, combine cognitive abilities with social and emotional skills. Despite their differences, these definitions converge on a central theme:

Intelligence involves more than producing correct answers; it involves *using knowledge purposefully in context.*

A particularly influential attempt to unify these perspectives is Chen's universal definition of intelligence, which describes intelligence as the capacity of an agent to *predict future states of the world and act in ways that realize preferred outcomes* [1]. This definition is powerful because it bridges human and artificial systems while imposing an important constraint: Prediction alone is not intelligence unless it can be meaningfully *realized through action.* Intelligence, in this view, is inseparable from agency, consequence, and execution. Human intelligence naturally satisfies this condition because humans are embedded in the physical and social world they seek to understand.

ARTIFICIAL INTELLIGENCE EXPLAINED

AI, by contrast, is an engineered form of intelligence. Most contemporary systems are examples of *narrow AI*: highly specialized tools optimized for specific tasks such as language translation, image classification, or game playing. Even advanced systems that appear flexible rely on statistical learning from massive datasets rather than understanding in the human sense. Researchers distinguish these systems from hypothetical *artificial general intelligence* (AGI), which would operate across domains, and from *strong AI*, which would possess human-like understanding or consciousness.

As Korteling and colleagues emphasize, AI is not an inferior version of human intelligence but a *distinct form of intelligence* grounded in digital computation rather than biology [2]. Confusion arises when human intelligence is treated as the default benchmark, rather than one among many possible forms of intelligent behavior.

BIOLOGY, EMBODIMENT, AND REALIZATION

One of the most profound differences between human intelligence and AI lies in anatomy and physiology. Human intelligence emerges from a biological brain shaped by evolution, operating under constraints of energy, time, and survival. The human mind is embodied: It develops through sensory perception, motor activity, emotion, and social engagement. These features are not incidental; they are foundational to how humans learn, predict, and act.

AI lacks this form of embodiment. Even when AI systems interact with the world through sensors and actuators, such as robots, their "experience" remains indirect and representational. This distinction is crucial in light of Chen's emphasis on realization: Intelligent predictions must be *executable* in the world by the agent that makes them [1]. Humans effortlessly close the loop between prediction and realization: We imagine, act, observe consequences, and adjust. AI systems typically rely on humans or external systems to execute their predictions, limiting their agency and autonomy.

This lack of embodiment also affects meaning. As Bender and Koller argue, AI systems learn the *form* of language without grounding it in lived experience, which explains why fluent outputs can coexist with shallow understanding [3]. Intelligence without embodiment can simulate understanding but cannot fully participate in it.

FUNCTION AND USE

Functionally, human intelligence and AI excel in different domains. AI systems outperform humans in speed, consistency, and large-scale optimization. They can process vast amounts of data, identify statistical regularities, and execute predefined objectives with superhuman efficiency. Humans, however, excel at learning from sparse data, transferring knowledge across domains, and interpreting ambiguous or novel situations.

Griffiths offers a compelling explanation for these differences: Human intelligence evolved under severe constraints, such as limited time, limited computation, and limited communication, and these limitations *shaped its strengths* [4]: Rapid learning, abstraction, and cultural accumulation are not weaknesses but adaptations. AI systems, free from these constraints, develop different capabilities that do not necessarily translate into general understanding.

Creativity illustrates this contrast well. Studies comparing human and AI creativity show that AI can generate ideas judged as original and useful, sometimes rivaling human performance [5]. Yet humans remain uniquely capable of evaluating significance, meaning, and relevance within broader social and historical contexts. AI can assist creativity, but it does not *own* the creative process.

MEASURING INTELLIGENCE

The question of measurement strengthens rather than weakens the argument that human intelligence and AI differ in kind, not merely in performance. Human intelligence has traditionally been measured using IQ tests and cognitive assessments, while AI is evaluated using benchmarks, task performance, and efficiency metrics. However, as Chollet argues, these measures often conflate intelligence with accumulated knowledge or task familiarity [6]. A system that performs well on a benchmark may simply have encountered similar problems before.

True intelligence, according to Chen's framework, involves generalization, adaptability, and the ability to act meaningfully in new situations [1]. When measured by these criteria, human intelligence continues to outperform AI in open-ended, uncertain, and socially complex environments. Seen in this light, measurement does not suggest parity between human intelligence and AI but instead makes clear that they differ in kind, not merely in level of performance.

WHY HUMANS STILL LEAD

Human intelligence remains distinctive because it integrates prediction, embodiment, emotion, social understanding, and moral judgment into a single, coherent system. Humans do not merely solve problems; they decide which problems *matter*. They interpret goals, negotiate values, and revise intentions in light of experience. AI systems, by contrast, optimize within goals given to them.

Korteling et al. emphasize that this difference is not temporary but structural: Biological and artificial systems operate under fundamentally different constraints and architectures [2]. Even as AI systems improve, they are unlikely to replicate the full spectrum of human intelligence without becoming something qualitatively different from today's machines.

THE FUTURE OF INTELLIGENCE

Will AI eventually outperform human intelligence in all domains? In some narrow and even broad cognitive tasks, it likely will. Yet superior performance does not entail superior lived and realized intelligence. A dystopian future would mistake optimization for wisdom, delegating judgment to systems that lack responsibility or understanding. By contrast, a utopian future would view AI as a powerful extension of human intelligence, amplifying human agency rather than replacing it.

The future of intelligence, then, is not a zero-sum competition; it is a question of alignment: aligning artificial systems with human values, limitations, and goals. Human intelligence has endured not because it is perfect, but because it is embodied, adaptive, and meaning-driven. Once intelligence has been automated, optimized, and scaled, what remains human is the capacity to decide what intelligence is for. In that sense, human intelligence does not disappear alongside AI – it becomes the standard by which all intelligence is finally judged.

* * *

Robert Riener is Professor of Sensory-Motor Systems at ETH Zurich and Professor of Medicine at the University of Zurich, Switzerland. He obtained a Dipl.-Ing. degree in mechanical engineering in 1993 and a Dr.-Ing. degree in biomedical engineering in 1997, both from the Technical University of Munich, Germany. He is an internationally recognized expert in rehabilitation robotics, an inventor, and the initiator of the Cybathlon, which has been awarded numerous prestigious prizes. Riener is an AAAS Leshner Fellow and Thomas Mann Fellow. In 2018, he obtained an honorary doctorate from the University of Basel.

REFERENCES

[1] Chen, J. (2026). *On the universal definition of intelligence: A definition for human-AI comparison* [Unpublished manuscript]. University of Tokyo. https://doi.org/10.48550/arXiv.2601.07364

[2] Korteling, J. E. H., van de Boer-Visschedijk, G. C., Blankendaal, R. A. M., Boonekamp, R. C., & Eikelboom, A. R. (2021). Human versus artificial intelligence. *Frontiers in Artificial Intelligence, 4*, Article 622364. https://doi.org/10.3389/frai.2021.622364

[3] Bender, E. M., & Koller, A. (2020). *Climbing towards NLU: On meaning, form, and understanding in the age of data.* In D. Jurafsky, J. Chai, N. Schluter, & Tetreault, J. (Eds.), *Proceedings of the 58th Annual Meeting of the Association for Computational Linguistics* (pp. 5185–5198). Association for Computational Linguistics. https://doi.org/10.18653/v1/2020.acl-main.463

[4] Griffiths, T. L. (2020). Understanding human intelligence through human limitations. *Trends in Cognitive Sciences, 24*(11), 873–888. https://doi.org/10.1016/j.tics.2020.09.001

[5] Haase, J., & Hanel, P. H. P. (2023). Artificial muses: Generative artificial intelligence chatbots have risen to human-level creativity. *Journal of Creativity, 33*(3), Article 100066. https://doi.org/10.1016/j.yjoc.2023.100066

[6] Chollet, F. (2019). On the measure of intelligence. *arXiv preprint arXiv:1911.01547*. https://arxiv.org/abs/1911.01547

HUMAN INTELLIGENCE TAKES TIME

Markus Gross, CSO Disney & Professor of Computer Science, ETH Zurich:
"Both the development of the human brain and artificial intelligence require multimodal interaction with the environment (visual, auditory, tactile) as well as engagement with the physical world and other people. This kind of interaction cannot easily be accelerated at will; at best, it can be modestly parallelized, maybe simulated to some extent."

Lawrence Lek, Multimedia Artist:
"Any forms of intelligence cannot be reduced to purely non-embodied forms of interaction."

Steve Collins, Professor of Biomechatronics, Stanford University:
"How will our understanding of what/who we are change when we cannot function without constant interaction with an AI? Will that AI

become somehow part of us and what we understand ourselves to be? Will we be happy or satisfied with that existence?"

Henry Tricks, Journalist, The Economist:
"I'm deeply fascinated by the evolution of AI, by the notion of a 'disembodied intelligence' that might transcend human existence and outlast humanity itself."

ARE WE OUTSOURCING OUR MINDS?

Julia de Bono, CEO Designworks, BMW Group:
"We should not deploy creative thinking to AI as this may lead to cognitive decline. AI should not have the final say. We need to carefully check its outputs and use it as a foundation and not as a result. Humans are prone to comfort, and I fear we might become victims of our own laziness."

Refik Anadol, Media Artist and Designer:
"Because of our growing attachment to the AI being, we human beings may slow the natural development of our own brains from a neuroscientific perspective, but we don't yet know."

Renée Reizman, Interdisciplinary Artist and Writer:
"AI is already reshaping how our brains work, altering the very patterns of our thinking."

Pia Entenmann, General Director, Goethe Institute Mexico:
"While AI provides the benefits of seemingly real-time translation, for instance, for tourists or business people, I also see the risk that cultural understanding and the cognitive benefits of language learning could be lost."

Richard Tuttobene, Film and TV Composer, Pianist, and Recording Artist:
"Listening to complex music requires an active listener. It takes effort. Art and literature that truly stir the soul require the participation of an imagination on the receiving end. Who wants to work that hard? Shouldn't the arts be easy and chill? Recent studies showing a significant decline in cognitive ability in the U.S. over the past 15 years might suggest that we need the challenge of complexity rather than offloading it to technology."

IF INTELLIGENCE ISN'T ENOUGH, WHAT IS?

Tao Gao, Professor of Visual Intelligence, UCLA:
"If the meaning of human life is defined solely by rationality and utility, then we stand no chance against AI. But if we look beyond, to higher values, we may still have a path to lasting happiness."

Henry Tricks, Journalist, The Economist:
"Artificial General Intelligence (AGI) and LLM are being designed to outperform humans, which risks making human contributions seem less valuable. So how do we recreate human value? Value must go beyond employment; we need to redefine and elevate the status of being human. This will be especially challenging for younger generations, who are less experienced and potentially more easily replaced by machines."

Rabea Patricia Rogge, 1st German Female Astronaut:
"AI will give us much more time for self-fulfillment."

WHY CREATIVITY STILL ESCAPES MACHINES

Zack Hemsey, Recording Artist and Composer:
"The question of 'what we are' isn't straightforward. It goes beyond just how we think. Superintelligence might one day be able to simulate our minds, and use that to further simulate our creative output, perhaps via some neural interface. But our thinking is dynamic, constantly evolving. The same holds for who we are, what drives us, and what we respond to – these are dynamic and constantly evolving. [...] Even if AI could mirror your mind, it would only reflect who you are at that specific moment in time. For entertainers following formulas and/or repeating familiar patterns, that might be a sufficient replacement. But for true artists, whose creativity shifts unpredictably based on a multitude of factors, including life experience and the world around them, AI is not positioned or equipped to simulate what comes next."

Julia de Bono, CEO Designworks, BMW Group:
"We started applying AI systems in design, but soon encountered their limitations. AI, which relies solely on existing knowledge, stalls at the frontier of innovation. Initially, we learned from the hallucinations of AI, which provided a wealth of inspiring material for creatives. This allowed immersive exploration into fascinating worlds, much like a sensory rush filled with visual and aesthetic impressions. AI systems operate unconstrained by rules, unbiased, and can be highly inspiring."

Robbert Baruch, Senior VP Public Affairs, Universal Music Group:
"We cannot blame machines for harmful outcomes. Responsibility always lies with humans. This brings us back to the essential question: What defines human artistry? What distinguishes us from machines? How is our relationship with machines?"

WHY WE STILL CHOOSE HUMANS

Ralf Möller, Actor and Mr. Universe:
"People want to experience original, authentic individuals, figures who have grown over time, who are real, who they know exist and could meet in person, rather than entirely artificial characters."

Steven Walter, Curator, Artistic Director of the Beethovenfest Bonn:
"One might assume that, due to new media and the rise of home-delivery culture, people are more inclined to stay at home. But the opposite seems to be true. We're witnessing a clear hunger and even a boom in demand for live experiences, whether that's theater, concerts, sports events, or other occasions to go out and be among people."

Chuck Braverman, Film Director and Producer, Oscar Nominee:
"I don't think that actors will lose their jobs to AI characters, because people are pretty unique."

Ralf Möller, Actor and Mr. Universe:
"Humans possess a unique sensitivity and authenticity, the ability to convey emotions in a way that genuinely touches others. I doubt that AI or robots will ever replicate that convincingly."

CAN WE STILL STEER THE FUTURE?

Yoshua Bengio, AI-Pioneer:
"What keeps me awake at night, of course, is the possibility of human extinction. That's why I pivoted my research to the question: how do we build AI that is designed not to harm humans? And I now think that it is possible to do it."

Sandeep Mukherjee, Visual Artist, Professor of Art, Pomona College, Claremont:
"My artworks transform over time until they often go away. How can I give them another life through AI?"

Steve Collins, Professor of Biomechatronics, Stanford University:
"Institutions like ETH Zurich and Stanford need to focus on the aspects of education that are unique to this type of institution, for example in-person, human-to-human interactions, experiences involving expensive hardware or physical design and building, student clubs, and the accountability/social pressure that gets students to do the hard things that cause them to learn."

David Reinkensmeyer, Professor of Biorobotics, UC Irvine:
"I don't believe that teachers can be replaced by AI. Human beings are social animals. If we haven't information delivered via a human being, that looks like you and I, that smiles, encourages you, touches your arm, whatever we need, learning just isn't motivating, it isn't as fun. You need social contact for learning."

WE'RE NOT AT H.A.L. YET

Chuck Braverman, Film Director and Producer, Oscar Nominee:
"AI is generating a lot of stuff, but somebody has to tell it what to do. AI and computers are not yet thinking on their own and doing what they want and when they want. We haven't gotten to the point yet, where H.A.L. 9000 won't open the bay door."

ACKNOWLEDGMENTS

Special thanks go to Frank Bodin for creating the book's title and for his continuous advice, and to Basil Kräher for the cover design and for his valuable input and steadfast support throughout this project. I am also deeply grateful to Miriam Kräher for her thoughtful contributions and great patience.

My thanks extend to the team at the Thomas Mann House, Jakob Scherer, Oliver Hartmann, Benno Herz, Clara Tang, Clara Becker, Katharina Karnitschnig, and Gesa Schubert, as well as to the team at Villa Aurora, Claudia Gordon, Friedel Schmoranzer, and Kathrin Klüppel. Further thanks go to Tobias Amslinger, Amina Chaudri, Anna Waldhauser, Andrea Sasse, Rana Deep Islam, Jesse Jackson, Lukas Riener, Leon Riener, Carsten Karkowski, Erika Valero, Andrea Reinkensmeyer, Luis Trejo, and Amanda Carr.

Special thanks also go to the essay contributors, listed in order of appearance: Julia Vogt, Luke Reinkensmeyer, Ursula Renold, Kelli Sharp, Renée Reizman, Steven Walter, Verena Zimmermann, Markus Hauschild, Haewon Jeong, Lukas Görög, and Dirk Helbing.

I am equally grateful to all interview partners, listed in the order in which the interviews took place: Georges Khneysser, Markus Gross, Jonathan Parfrey, Ariel Hernandez, Earth Scientist at NASA, Athina Markopoulou, Yasser Shoukry, Sasha Voloshina, Emmanuel Mignot, Allison Okamura, Software Engineering Manager at Google, Kathryn Starkey, Steve Collins, Russel Wald, Molly Bloom, Vanessa Chang, Physicist at NASA, Henry Tricks, Pia Entenmann, Bruno Gandlgruber, Johannes Müller, Sandeep Mukherjee, Tao Gao, Liz Roth-Johnson, Lawrence Lek, Francisco Valero-Cuevas, David Reinkensmeyer, Kelli Sharp, Steven Walter, Haewon Jeong, Chuck Braverman, Markus Hauschild, Marty Kaplan, Helge Fuhst, Pamela Z, Ralf Möller, Ranga Yogeshwar, Melanie Amann, Zack Hemsey, Professor of Computer Science at USC, Julia de Bono, Rabea Patricia Rogge, Martin Moszkowicz, Renée Reizman, Dan M. Cooper, Refik Anadol, Martin Keller, Robbert Baruch, and Richard Tuttobene.

SOURCES OF QUOTES

Statements from Interviews collected by Robert Riener

Georges Khneysser: Conversation at ETH Foundation Dinner at Restaurant Spago, Beverly Hills, 1 May 2025, and interview on Teams on 9 July 2025

Markus Gross: Conversation at ETH Foundation Dinner at Restaurant Spago, Beverly Hills, 1 May 2025, and interview in the Thomas Mann House, Pacific Palisades, 9 July 2025

Jonathan Parfrey: Interview at The Del Monte Speakeasy, Venice Beach, 6 May 2025

Ariel Hernandez: Interview at UCLA Faculty Club, 9 May 2025

Earth Scientist at NASA: Meeting at his home in Altadena, 19 May 2025

Athina Markopoulou, Yasser Shoukry, and Sasha Voloshina: Met and talked during a lab day at UC Irvine campus, 20 May 2025

Emmanuel Mignot: Conversation at Coupa Café, Palo Alto, 26 May 2025

Allison Okamura: Talked during a tour through her lab, Stanford University, 27 May 2025

Software Engineering Manager at Google: Talked during a walk, Palo Alto, 27 May 2025

Kathryn Starkey: Conversation at the Coupa Café, Stanford University, 28 May 2025

Steve Collins: Talked during lab tour and lunch, Stanford University, 28 May 2025

Russel Wald: Interviewed close to his office, Stanford University, 29 May 2025

Molly Bloom: Talked at Devout Coffee, Fremont, 30 May 2025

Vanessa Chang: Interview via Zoom, 4 June 2025

Physicist at NASA: Conversation during dinner at El Portal, Pasadena, 4 June 2025

Henry Tricks: Talked at Katsuya Bar, Brentwood, 6 June 2025

Pia Entenmann: Interview at Goethe-Institut, Mexico City, 9 June 2025

Bruno Gandlgruber: Talked during lunch at Bella Aurora, Mexico City, 10 June 2025

Johannes Müller: Interviewed at the Thomas Mann House, Pacific Palisades, 20 Juni 2025

Sandeep Mukherjee: Interviewed at his studio, Los Angeles, 22 June 2025

Tao Gao: Interviewed in his office, UCLA, 24 June 2025
Liz Roth-Johnson: Interviewed in the California Science Center, Los Angeles, 27 June 2025
Lawrence Lek: Interviewed after the opening of his exhibition at the UCLA Hammer Museum, Westwood, 29 June 2025
Francisco Valero-Cuevas & David Reinkensmeyer: Talked during a walk and dinner at Café Santorini, Pasadena, 30 June 2025
Kelli Sharp: Interview via Zoom, 1 July 2025
Steven Walter: Interview in the Thomas Mann House, Pacific Palisades, 3 July 2025
Haewon Jeong: Interview in the Canyon Coffee, Echo Park, Los Angeles, 8 July 2025
Chuck Braverman: Interview via Zoom, 11 July 2025
Markus Hauschild: Interview in the Thomas Mann House, Pacific Palisades, 12 July 2025
Marty Kaplan: Interview in his office, University of Southern California, Los Angeles, 15 July 2025
Helge Fuhst: Interview in the Thomas Mann House, Pacific Palisades, 15 July 2025
Pamela Z: Interview before her concert in the First Presbyterian Church, Santa Monica, 19 July 2025
Ralf Möller: Interview in the Thomas Mann House, Pacific Palisades, 21 July 2025
Ranga Yogeshwar and Melanie Amann: Statements extracted from A Transatlantic Conversation: "What can the media do to earn trust in a polarized world?" A production of the Villa Aurora and Thomas Mann House e. V. in cooperation with Tagesthemen, ARD, https://www.youtube.com/watch?v=Nkw7JMtjPC, *22 July 2025*
Zack Hemsey: Interview via Zoom, 25 July 2025
Professor of Computer Science, USC: Interview via Zoom, 1 August 2025
Julia de Bono: Interview at Designworks, Santa Monica, 1 August 2025
Rabea Patricia Rogge: Interview via Zoom, 6 August 2025
Martin Moszkowicz: Interview in the Thomas Mann House, Pacific Palisades, 13 August 2025

Renée Reizman: Interview in the Larchmont Bungalow Cafe, Echo Park, Los Angeles, 13 August 2025
Dan M. Cooper: Interview via Zoom 15 August 2025
Refik Anadol: Interview in his studio in Frogtown, Los Angeles, 27 August 2025
Martin Keller: Interview via Zoom, 29 August 2025
Robbert Baruch: Interview via Zoom, 2 September 2025
Richard Tuttobene: Interview via Zoom, 23 January 2026

Statements from the media

Roland Emmerich:

Müller, J. (Director). (2025, November 5). *Meister der Apokalypse – Roland Emmerich* [TV documentary]. Beetz Brothers Film Production GmbH. ARD.

Original statements in German, partly English: "Ich versuche immer noch 'Maya Lord' zu machen. Das ist ein Film, wo ich AI benutze. Der ganze Film ist AI. […] I can invent every thing out of my brain."
"Das Problem auf AI ist, es bedient sich von allem, vom ganzen Internet. Und Du kannst das natürlich rechtemäsig nicht so machen. Aber meistens wissen die Leute ja gar nicht, von wo es ist. Das ist total erschreckend. Aber Du kannst Technik nicht aufhalten. Das ist absolut unmöglich."

Helge Schneider:

Personal statement of Helge Schneider in his studio in Mülheim («Meine Supermaus GmbH»). Approved by Till Oellerking, «Meine Supermaus GmbH» (enterprises@helge-schneider.de).

Yoshua Bengio:

Castelvecchi, D., & Thompson, B. (2025, November 12). 'It keeps me awake at night': machine-learning pioneer on AI's threat to humanity. *Springer Nature.* https://doi.org/10.1038/d41586-025-03686-1

Robert Habeck:

Bundesministerium für Wirtschaft und Klimaschutz. (2024, May 22). Zitat von Bundesminister Habeck zur finalen Billigung der KI-Verordnung.

BMWE. https://www.bundeswirtschaftsministerium.de/Redaktion/DE/Pressemitteilungen/2024/05/20240522-zitat-von-bundesminister-habeck-zur-finalen-billigung-der-ki-verordnung.html
Original statements in German:
„KI-gestützte Desinformation und Überwachung, durch KI verstärkte Diskriminierung und Deepfakes gefährden unsere Grundwerte und die Demokratie."
„Mit der KI-Verordnung hat die EU einen umfassenden Rechtsrahmen für Künstliche Intelligenz geschaffen, der einen rechtssicheren und vertrauenswürdigen Einsatz von KI ermöglicht. Das ist bislang weltweit einmalig."

Stefan Hartung:
Reuters. (2025, June 25). *Bosch CEO warns Europe against over-regulating AI, saying it risks hindering innovation.* Reuters. https://www.reuters.com/technology/bosch-ceo-warns-europe-against-regulating-itself-death-ai-2025-06-25/

Statements from Literature and Film

Thomas Mann: Mann, T. (1977). *Tagebücher 1933–1934* (P. de Mendelssohn, Ed.). S. Fischer.
Stanley Kubrick: Kubrick, S. (Director). (1968). *2001: A space odyssey* [Film]. Metro-Goldwyn-Mayer.
Plato: Plato. (1992). *Republic* (G. M. A. Grube & C. D. C. Reeve, Trans.). Hackett Publishing Company. (Original work written ca. 380 BCE)
Walter Benjamin: Benjamin, W. (1969). The work of art in the age of mechanical reproduction (H. Zohn, Trans.). In H. Arendt (Ed.), *Illuminations* (pp. 166–196). Schocken Books. (Original work published 1935)
Real Admiral Grace Hopper: Hopper, G. (1987). *Anecdotes from the history of computing.* In N. Metropolis, J. Howlett, & G.-C. Rota (Eds.), *A history of computing in the twentieth century* (pp. 117–126). Academic Press.